ROLE OF THE
DESIGN PROFESSIONAL
IN SOCIETY

Edited by William A. Kitch

ROLE OF THE
DESIGN PROFESSIONAL IN SOCIETY

Edited by William A. Kitch

California State Polytechnic University - Pomona

University Readers™
San Diego, CA

First published in the United States of America in 2010 by University Readers, Inc.

Trademark Notice: Product or corporate names may be trademarks or registered trademarks, and are used only for identification and explanation without intent to infringe.

14 13 12 11 10   1 2 3 4 5

Printed in the United States of America

ISBN: 978-1935551-95-9

University Readers™
800.200.3908 | www.universityreaders.com

Contents

PART THREE – Contracts

PART ONE

Articles

1: Project Delivery and Selection

By William A. Kitch

How do major projects actually get designed, created, and delivered to the client or user? How are the designers and the fabricators or builders selected? What is the role of the design professional in this process? In order to answer the last question we must first address the preceding two questions. The way the project delivery is structured defines the playing field on which the design professional will play. The selection process lays down the rules of the game. Without understanding the playing field and the rules of the game, you will be a spectator in the stands rather than a player on the field.

The goal of this article is provide structure and language to describe the different ways projects are delivered and designers and constructors are selected. In truth there are a myriad of methods for both delivery and selection and many different terms used to label the various methods. In this article we will develop a simple but useful model for delivery and selection. We will also build a common language for describing the methods. As you start your professional career, you will run into different models and different terms for project delivery and selection; but if we do a good job in this article, you will have a solid base to understand all the variations and flavors that currently exist in practice.

1.1. Project Phases

Whatever the project, a bridge, a home, a video game console, or a golf course, the project begins with an idea or a need perceived by the user. It is a long road from the user's first perception of a need, to the delivery of the final project.

Delivering major projects is a process that can take years or even decades. There are many steps in the process and many different persons involved. Nobody ever woke up in the morning and said "Hey, I think I'd like to build a 27,000 square foot warehouse today," and then went and hired a contractor to build the warehouse.

Most users are not design professionals and are therefore not able to determine the best way to meet the need they have identified. The users' initial statement of need is often very simple:

- We need to reduce congestion on the 10 freeway between the 57 and the 605.
- We need a warehouse that will meet our business demands for the next 15 years.
- I want a larger house.

It is not possible to go from identification of a need directly to building a solution unless the need is so simple and common that solutions already exist and can be bought off the shelf. If you have a very simple need such as storage for a lawnmower, then you can buy a prefabricated storage shed and erect it in your backyard. However, for needs even slightly more complex than this, there are several steps between identifying the need and building the solution.

For any significant project it is not possible to go directly from need to solution. What, then, is the first step after identification of a need? Many people would answer design. This may sound like a reasonable approach; identify your need, design a solution, and then you're ready to build the solution. However, this is usually a recipe for an expensive solution that doesn't really meet the user's need.

To understand why one cannot jump directly from a need to design, imagine you are an architectural & engineering designer and a user comes to you and says "We need a warehouse that will meet our business demands for the next 15 years." Would you immediately start designing the structure? Would your staff start sizing the beams and columns, designing the façade, and lay out the parking lot? No! And why not? Because you still don't have the input needed to start designing. You don't know how big the warehouse needs to be. You don't know if it needs office space and if so how much. You don't know the user's aesthetic requirements. Will they be happy with a tilt-up structure or do they want something more architecturally pleasing? Do they want to pay for a facility that will last for 75 years or are they happy to use it for 20 years and move on when it needs major repair or refurbishment? How many trucks will need to be loaded at one time? How much parking is needed?

To put it another way, the designer cannot start designing until she has the specific requirements for the design. In this example, the required square footage, amount of truck traffic, material to be stored, aesthetic requirements, and expected life span are all required before design can start. So there is a step between identifying the need and starting design. We will call that step planning.

We can now define the four phases that every project will go through: Planning, Design, Construction, and Use (or Occupancy). Traditionally, in the construction industry these phases have been accomplished in a simple linear process as shown in Figure 1.1a.

1.1.1. Planning

The initial phase of project delivery is the planning phase. The objectives of this phase are to turn the user's identified need into a clearly defined set of requirements, develop a program cost estimate, and identify funding resources.

The first two objectives, requirements definition and program cost estimation, normally require the services of engineers. Continuing with our warehouse example, the requirements definition would start by gathering data about the project. The designer would need to determine the square footage required for the warehouse. This information would come from the business records of the company and industry estimate of the amount of storage space required for the material being stored. The user's operations requirements would dictate the number of loading docks required, as well as the required parking spaces. A preliminary site investigation may be required to determine whether the structure can be built on a shallow foundation or if it will require a deep foundation. The goal at this point is to gather all information needed to develop a general scope for the project and identify the major cost items.

Once the general scope is determined (square footage of warehouse, office, parking requirements) the designer can develop options for the user. Options may include different sites, different sizes, or different quality of facilities. For each option, a program cost must be developed.

The program cost is essentially a budgetary cost. It is generally based on historic construction cost data and broad measures of the project scope such as square feet of warehouse space, square feet of pavement, or linear feet of sidewalk. The purpose of the program cost is to provide enough financial information to allow the owner to evaluate alternatives and to budget funds for the project. Program level cost estimates should include total project costs, including construction management, permitting, and other non-construction costs. Generally, program costs are satisfactory if they are within 20 to 30 percent of the final project costs.

Once options have been developed with program costs, the user is able to make a business decision about the best approach to meeting the identified need. In our warehouse example, the user may end up having to choose between two sites. The first may be close to her current location, thus providing low operational costs, but may provide a limited area with no expansion capability. The alternative site may be further away with lower construction costs and expansion capability, but have increased operational costs due to the distance from her current operations. For both options the designer should provide the scope and cost data needed for the user to evaluate and select between the options.

The planning phase is complete when the user has identified a general scope of the project, defined the design requirements, determined the overall expected project costs, and identified potential funding sources.

There are many roles for design professionals to play during the planning phase. Designers and planners are needed to translate the user's need into specific design requirements, to develop cost estimates, and to evaluate the alternatives. Some users, such as large utility companies, or state departments of transportation, may have the in-house engineering capability to complete the planning function. Other users, such as retail sales companies or small municipalities, may have to contract for the needed engineering and planning services.

1.1.2. Design

When the project scope and the design requirements are set, project design can start. The objective of the design phase is to develop the plans and specifications needed to build the project. The level of design detail needed will depend on the project delivery method. The traditional linear system shown in Figure 1.1a generally requires a complete set of detailed plans and specifications. The complete plans and specs as well as a construction cost estimate are needed before proceeding to the construction phase.

In the fast track process, shown in Figure 1.1b, it is often possible to reduce the level of detail required during design. This can provide a significant cost savings to the users. The differences between the linear and fast track processes will be discussed further in Section 1.2.

The design phase is, perhaps, the area most often associated with design professionals. This certainly is the phase of the project requiring the most extensive use of design skills and personnel. The breadth of design skills needed will vary greatly with type of project. An industrial project, such as our warehouse example, will require relatively few architectural skills and mostly engineering skills. The design of an upscale office park, in contrast, may require much more architectural design. Subdivision design makes extensive use of urban planners, civil and transportation engineers, and little architectural design.

1.1.3. Construction

The vast majority of work done during construction is the actual craft and trade work required to physically complete the project. However there is significant use of design professionals in this phase. The project designers will complete all of the design engineering work needed to prepare the final plans and specifications. However, there is still much design work needed beyond what is included in the plans and specifications. For example, temporary structures required only

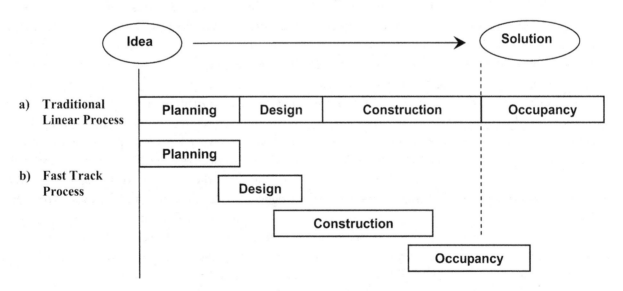

Figure 1.1: Project Phasing

for construction such as false work or shoring must be designed. Construction contractors often require design services to prepare alternative specifications or to develop specific construction techniques.

There is extensive communication between the designer and constructor during this phase. It is common for the designer to have a resident engineer on site at the construction project. The role of this resident engineer is to communicate between the construction contractor and the designer. The resident engineer provides the designer and owner information about the progress of construction and provides the construction contractor with interpretations of the plans and specifications from the designer.

Another function required during the construction phase is construction management. This role is often filled by an engineer.

1.1.4. Occupancy

While some may think there is little role for the engineer during the occupancy phase this is not true. The amount of engineering services needed during occupancy will vary significantly with the type of project, but in nearly all cases there is a significant role for engineering. Facility or maintenance engineering is required during the life of most projects in order to keep the project or facility in good operating order and to minimize the life cycle costs of owning the facility. This is true of roads, bridges, utility systems, and buildings. During the initial occupancy there is a lot of communication amongst the designer, the constructor, and the facility or maintenance engineer. The maintenance staff may require training to properly operate mechanical and electrical systems. Preventative maintenance programs must be developed often with the input of the designers. There will often be discrepancies between the design specifications and the operational capabilities. It normally requires engineering analysis to determine if the discrepancies are due to design deficiencies or construction problems.

It is useful to distinguish among the planning, design, construction, and occupancy phase, but it is often difficult to draw clear boundaries between them. Properly completing each phase requires a mix of skills, but engineering and design skills are required in all phases of project delivery.

1.2. Project Delivery Methods

As discussed earlier, the project delivery process requires many different skills including finance, costing, designing, engineering and planning. Each phase of the process has distinct goals and requires a special mix of skills. In all cases it requires a team or teams of people to complete a project. How these teams are selected, organized, and managed has significant impacts on how the project will proceed. We will use the term "delivery method" to describe the combination of contract organization and contractor selection that creates the organization of teams responsible for planning, designing, constructing, and delivering a project.

There are two critical variables that determine the type of delivery method. These are the contract structure (number of contracts) and the process used to select designers and constructors. By evaluating these two variables we are able to classify the most common delivery methods.

1.2.1. Contract Structure

Usually, the user or project owner will have to contract for both the design and construction services needed to complete a project. In some cases an owner may have in-house design resources. For example, Caltrans has extensive design capability and performs design services for many (but not all) of their projects with in-house resources.

In most cases the owner is contracting for both design and construction services. There are two basic ways to organize the contracts. In the traditional method, the owner lets two separate contracts: one for design services and one for project construction. This contract structure is shown in Figure 1.2a. The designer completes the design and provides complete plans and specs for the project. This structure gives the owner the most control over the design process but provides the least amount of collaboration between the designer and the constructor.

In the design-build process, the owner lets a single contract for both design and construction services.

This contract structure is shown in Figure 1.2b. This structure provides for the best collaboration between designer and constructor and has the potential for reducing both total time and cost of the project. However, this method gives the owner the least control over the process.

1.2.2. Selection Methods

There are many different methods for selecting design and construction services. However, most of these methods can be classified into one of three categories: low bid, best value, and qualification based selection.

1.2.2.1. Low Bid

The distinguishing characteristic of the low bid selection process is that selection is based strictly on price. There are different ways to specify the price. The price may be lump sum, a single price for all services, or it may be based on unit pricing. In unit pricing, rather than specifying the total cost of a project, the price is divided into areas and payment for each area is based on the unit prices. For example, base course placement may be priced in cubic feet of in-place compacted base course; painting may be priced in square feet of walls painted. Regardless of how the price is structured, this selection process is based solely on price.

1.2.2.2. Best Value

The distinguishing characteristics of the best value process are that the selection is made using both costs and some measure of the qualifications or quality of services provided by the bidder. Since it is not possible to put a monetary value on the qualifications of the bidders, this method uses some sort of weighting system to evaluate both qualifications and price. The weight systems used vary greatly.

Although there is no standard weighting system, a typical best value method is called the two step or two envelope method. In this method bidders provide separate qualifications and cost proposals. The qualifications are evaluated first by the owner and each competing firm is given a score based on the firm's qualifications. Next the cost proposals are evaluated. The owner then uses the cost proposal and the qualification rating to determine which firm is the best value to the owner. Conceptually, this method gives the owner the ability to pay more for a higher quality firm. However, because it isn't possible to put an exact cost on a given level of quality, it is difficult to determine and defend the trade off between cost and quality.

1.2.2.3. Qualifications Based Selection (QBS)

The distinguishing characteristic of the QBS process is that cost is not a selection criterion. In this process the selection is based solely on the bidders' qualifications. The price is negotiated after the selection is made. The process follows these general steps:

1. Owner defines general scope of work and publishes a request for qualifications.

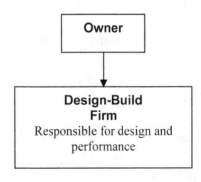

a) Separate contracts   b) Single contract

Figure 1.2: Contract Structures

2. Interested firms prepare proposals listing their qualifications, similar projects completed, and approach to the proposed project.

3. Owner evaluates qualifications submitted and selects the most qualified firm.

4. Owner and selected firm jointly define scope and cost of project. This is where the price/cost trade-offs are made.

5. Once cost and scope are agreed upon, a contract for the project is signed.

6. If owner and selected firm are unable to agree upon cost and scope, owner moves on to next best qualified bidder and attempts to negotiate cost and scope.

1.3. Defining Costs

What does a project cost? The answer to this question depends a lot on how you ask the question and to whom? In order to be able to compare apples with apples, we must have a common definition of costs. We will use the following definition from the Georgia State Financing and Investment Commission (1991).

The total construction cost (TCC) is the sum of
- Construction Cost of Work: Material and labor required to build the project
- General Conditions: Non material and labor costs such as permitting, leases for site offices and staging areas etc.
- Contractor's Fee: Contractor's profit

The design fees are the total cost paid for the design of the project.

The total design and construction cost (TDCC) is the sum of
- Total construction costs
- Design fees

Total Project Cost (TPC) is the sum of
- Total construction costs (TCC)
- Design fees
- Balance of project costs: Costs such as construction management fees, cost of capital, owner's in-house management costs, etc.

When comparing costs it is critical to understand if a particular cost quote is the TCC, TDCC, or TPC.

Ultimately, the cost that should concern the owner is the total cost of ownership (TCO) of a facility. This cost is the sum of the TPC plus all the operations and maintenance (O&M) costs incurred over the life of the facility. O&M costs far exceed TPC. O&M costs often make up 80% or more of the TCO.

The O&M can be strongly impacted by the design. It is often true that one can decrease the TCO by spending more on the TPC. For example making a building more efficient may increase TPC but savings in O&M costs will far outweigh the increased TPC and reduce the TCO. Unfortunately TCO is often not considered in the project delivery selection process.

1.4. Defining Delivery Methods

Using the variables of contract structure and selection method we can define six different project delivery methods. Table 1.1 identifies the various delivery methods based on the number of contracts and selection process. Each delivery method will be discussed in further detail.

1.4.1. Design-Bid-Build

Defining Characteristics: Separate contracts for designer and builder. Builder selected solely on cost.

Other Typical Characteristics: The designer generally completes 100% design before the build contract is let. Communication between designer and builder is through the owner.

The design-bid-build process is shown in Figure 1.3. The delivery method provides for the least interaction between the designer and the builder. It is not amenable to fast tracking. It does provide for the highest level of cost competition for the construction phase since cost is the only selection criteria. This method can be very successful for the owner for projects with a well-defined scope that use well established construction methods, and do not require or benefit from new or innovative processes.

Table 1.1: Project Delivery Methods

| Selection Process | Contract Structure | |
	Two Contracts Designer & Builder	One Contract Designer-Builder
Low Bid	Design-Bid-Build	Design/Build Low bid
Best Value	CM/GC Best Value	Design/Build Best Value
Qualifications Based Selection	CM/GC QBS	Design/Build QBS

1.4.2. Construction Manager/General Contractor with Best Value Selection (CM/GC Best Value)

Defining Characteristics: Separate contracts for designer and builder. Builder selected on qualifications and TCC.

Other Typical Characteristics: Designer generally completes a preliminary design before build contract is let. Fast tracking is often used. Collaboration between designer and builder determines final design and cost.

The CM/GC process is shown in Figure 1.4. The delivery method generally provides for more interaction between the designer and the builder because the design is not complete when the builder is selected and the owner needs to build a collaborative process with the designer and builder to determine final cost and scope. This process is amenable to fast tracking but still requires separate procurement time for both the design and construction contracts. The cost portion of the evaluation in this process is usually a guaranteed maximum TCC based on the preliminary design. The final TCC is determined during the design and negotiation process. In order to be able to accurately compare costs, the preliminary design must be complete enough that potential construction bidders can develop meaningful TCC estimates.

1.4.3. Construction Manager/General Contractor with Qualifications Based Selection (CM/GC QBS)

Defining Characteristics: Separate contracts for designer and builder. Builder selected solely on qualifications.

Other Typical Characteristics: Designer generally completes a preliminary design before build contract is let. Fast tracking is often used. Collaboration between designer and builder determines final design and cost.

The process is the same as the CM/CG Best Value method (Figure 1.4.). The main difference between the two methods is that this method does not require a cost proposal from potential builders since the QBS process is based solely on qualifications. This means that the preliminary design only needs to have a general scope of the project. The final project scope and cost will be negotiated after selecting the builder.

Note that in both of the CM/GC delivery methods any cost savings realized in the design and scope development phase go to the owner. While this potentially provides the owner with a great deal of value it does not incentivize the builder into taking risks that could have potential cost savings since the builder would not benefit from those savings.

1.4.4. Design-Build Low Bid

Defining Characteristic: A single contract for both design and construction. Selection based solely on TDCC.

Other Typical Characteristics: Generally includes some sort of fast track organization. Pricing usually based on a guaranteed maximum TDCC.

The design-build process streamlines the total project timeline by eliminating one contract procurement since both the design and construction services are combined in a single contract, as shown in Figure 1.5. It also maximizes collaboration between designer and builder since they are on the same team, under a single manager. The low bid selection process maximizes cost competition among bidders. However

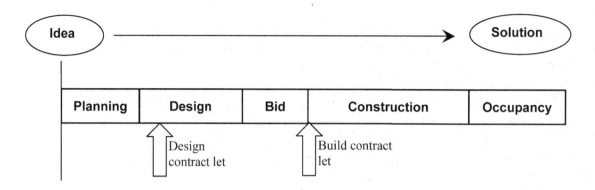

Figure 1.3: Design-Bid-Build Process

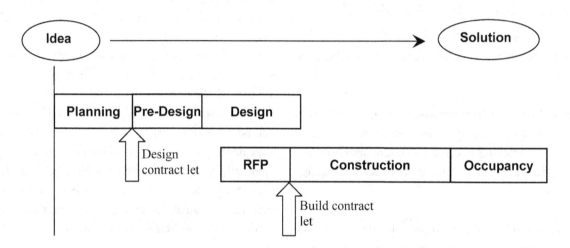

Figure 1.4: CM/GC Process

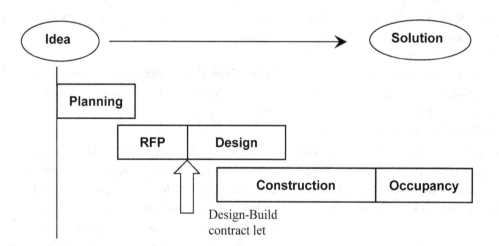

Figure 1.5: Design-Build Process

all of the savings a contractor might realize with an innovative design-build idea are kept by the contractor. The low bid selection method does not allow the owner to make cost-schedule-scope trade-offs since the selection is based solely on cost.

1.4.5. Design-Build Best Value

Defining Characteristics: A single contract for both design and construction. Selection based on both qualifications and TDCC.

Other Typical Characteristics: Generally includes some sort of fast track organization. Pricing usually based on a guaranteed maximum TDCC.

The process is the same as other design-build delivery methods, see Figure 1.5. Like other design-build methods, it maximizes collaboration between designer and builder. The best-value selection process allows the owner some control of the cost-scope trade-off, since all bidders will generally have to provide their design-build approach in the qualifications part of the proposal.

1.4.6. Design-Build QBS

Defining Characteristics: A single contract for both design and construction. Selection based solely on qualifications.

Other Typical Characteristics: Generally includes some sort of fast track organization.

The process is the same as other design-build delivery methods, see Figure 1.5. Like other design-build methods, it maximizes collaboration between designer and builder. The QBS selection process allows the owner maximum control of the cost/scope trade-off since that negotiation occurs after selection of the design-build contractor. This delivery method requires the greatest sophistication on the part of the owner to both select and negotiate with the design-builder. Many owners use separate contracts with design professionals to aid in the selection process.

1.5. Selecting the Best Project Delivery Method

None of the project delivery methods is inherently better than another. The selection of a project delivery method will depend on a number of variables including: policies of owner's organization, in-house engineering capabilities, project timeline, and funding availability, just to name a few. What is important is for owners, designers and builders to understand the different characteristics of each method so that the project delivery method can be carefully matched to a specific project.

1.6. The Design Professional's Role in Different Project Delivery Methods

Design-bid-build has been the traditional method of project delivery and most design professionals are familiar and comfortable with this method. In this process the design professional has played one of two roles: the project manager for the owner (either on the owner's staff or under separate contract) or as the contract designer. As the contract designer, operating under a design only contract, design professionals have been used to working directly with the owner with little direct interaction with builders. The design-build models are quickly changing this relationship. In most design-build contracts the prime contract is the builder, and the designer is generally a subcontractor to, or minor partner of, the builder. This new relationship has proved difficult for many design professionals.

Some designers complain about the loss of authority that comes from working for the builder rather than the owner. When a designer works for the builder, the designer's primary responsibility is to the builder not the owner. Many designers struggle to balance the responsibility to produce a low cost design for the builder while meeting their professional standard of care. Whatever role the design professional plays, it is critical that the designer clearly understands the nature of the relationship that is being undertaken before that designer agrees to take on the work.

There is no reason for designers to avoid working under design-build delivery systems rather than the traditional design-bid-build systems they are familiar with. The designer simply needs to understand their role in the delivery system, and how to best apply their professional skills and standards to the project at hand.

References

Georgia State Financing and Investment Commission (2001), Selecting the Appropriate Project Delivery Method <http://dlg.galileo.usg.edu/ggp/id:s-ga-bf500-b-pm1-b2001-bp75-belec-p-btext> (Dec. 12, 2007)

Associated General Contractors of America and National Association of State Facilities Administrators (2004), Best Practices For Use of Best Value Selections, <http://qic.tbr.state.tn.us/Documents/Best_Value_06_07_04.doc> (Dec. 20, 2007)

American Institute of Architects—Illinois (2000), Qualifications-Based Selection A Guide Including Model Local Government Policy and Procedures for Selecting Architects, Engineers and Land Surveyors <http://www.acec-il.org/qbsguide.pdf> (Dec. 22, 2007)

Architects and Engineers Conference Committee of California (1993) Qualifications-Based Selection: A Guide for the Selection of Professional Consultant Services for Public Owners, <http://www.cspe.com/FunctionalAreas/GovtAffairs/QBSManual.pdf> (Dec. 22, 2007)

2: What is Law

By John W. Tulac

*L*aw is the framework for any civilized society (any uncivilized society too, but we don't want to go there).

Law covers just about everything that we do. We are presumed to know the law. Go to the library. You'll find several rows of law books. You are presumed to know everything contained in those books and they are only a very small part of the total body of law. Remember …

IGNORANCE OF THE LAW IS NO EXCUSE!

Law is not for the lawyers. Law is for everyone. Anyone can learn the basics of law relevant to everyday business and everyday life.

Jurisprudence is concerned with the philosophy of law. There are several schools of thought and the study of jurisprudence is endlessly fascinating. It is also unnecessary to our needs. For our purposes, law defines what is right or wrong to do in our business and everyday lives. Law sets a minimum threshold of conduct. To fall below the threshold is to engage in unlawful conduct. Such unlawful conduct may be a civil wrong (a tort or a breach of contract) or it may be a criminal wrong (an act that is also deemed to harm not only the victim but against society as well).

In the United States, there are four sources of law. The primary source is the United States Constitution. In a jurisprudential sense "We the People" are the ultimate source of law as embodied in the Constitution. The other sources of law are statutory law, case law and common law, and administrative or regulatory law. All four sources of law impact what we do as businesspeople and citizens.

Exercise: When you think of law, what words come to mind? List as many of them as you can. Give yourself fifteen minutes. List them here:

Most people will associate law with justice. However, not all laws are just. People also disagree about what might be just under a given set of circumstances.

People also associate law with fairness. Law and fairness often overlap but they are not the same things. In contract law, the fairness of a contract is generally irrelevant. This isn't as difficult to accept as it might first seem. A toddler soon learns to add "that's not fair" to his list of most used and useful vocabulary. However, what one comes to consider fair changes as one grows up (of course, some people never grow up and may need to be handled like toddlers).

Through the use of law, we hope to achieve objectively fair and just results. And hell is paved with good intentions. Many bad laws have been passed in the name of worthy causes. Many had unexpected bad consequences. Visit dumblaws.com for a hilarious collection of mostly well-intended but nevertheless bad or just plain stupid laws. Here are a few examples:

- No vehicle without a driver may exceed 60 miles per hour.
- It is illegal to wear a fake moustache that causes laughter in church.

- If two trains meet on the same track, neither shall proceed until the other has passed.
- It is mandatory for a motorist with criminal intentions to stop at the city limits and telephone the chief of police as he is entering the town.

This course is not an academic exercise in the law. It is designed to give you minimum competence in recognizing legal issues. The focus in presenting the law is on what is practical. You cannot learn everything you need to know in ten short weeks. You <u>can</u> learn enough to recognize legal issues and to develop an approach to prevent legal issues from becoming legal problems. You will also learn a methodology to help you think clearly and logically about legal issues.

Law, then, is for everybody. It is in your own best interests to know the law that is relevant to what you do.

2.1. Law and the Design Professional

What does it mean to be a professional? As a licensed engineer or architect you have some things in common with all professionals including attorneys.

The law holds professionals to a higher duty or legal responsibility than ordinary citizens for anything that is specific or specialized to the professional's expertise or practice. You will learn more about the professional standard of care, which defines or measures this professional duty.

The law also considers a professional to be a fiduciary. A fiduciary is subject to the highest standard of care that the law requires of anyone. A design professional owes an obligation not only to the client but also to the public at large. The design professional must not only provide the client with what it wants, but must also advise the client what it needs as well. A professional provides services not only under contract to the client, but with a broader responsibility to the public as well. For a design professional, that broader responsibility includes a commitment to public safety.

A professional is licensed. This means that the state reserves the right to set the standards for admission to the profession and confer the right of the individual to be known as a professional. A professional is subject to a code of ethics. Everyone has an obligation to be ethical. A professional, however, is subject to specific rules of conduct that are codified and must be followed for the professional to remain in good standing.

2.2. Structure of Law and Legal Analysis

Law covers virtually every subject matter. Law is organized into many categories. In this class we deal primarily with the categories of tort and contract law.

Human beings are held accountable for their actions and sometimes their failures to act. Conduct is examined under the law pursuant to objective standards. The law does not require perfection, merely what is reasonable under the circumstances. The law requires us to act in certain ways when we do act and sometimes requires us to act even though we may not want to act. How do we know when and how to act? We must learn the applicable *standard of care* required under any given circumstances. We will spend a lot of time on this concept.

Unfortunately, people do things that injure one another or which damage or destroy property. Even an uncivilized society has rules to deal with such things. In our society we recognize that a wrong can be civil, criminal or both.

A civil wrong is called a *tort*. It is a wrong that harms an individual or property. A *crime* is a wrong against society as a whole. Not all crimes are torts and not all torts are crimes. Examples: Defamation, which is a false statement about another that puts that person in a bad light, is a tort, but not a crime. Prostitution, an exchange of a sexual act for cash or other consideration, is a crime, but not a tort. A battery, which is a harmful or offensive touching of another without consent, is both a tort and a crime.

A *claim* is an allegation of wrongdoing that has caused a harm or loss. A *cause of action* is a claim recognized under the law for which a remedy is available. A cause of action consists of *elements,* each of which must be proven or the cause of action fails. You must learn how to recognize causes of action and their elements and how to prove or disprove each element using the facts known or inferred.

2.3. IRAC Methodology

An essay question allows you to communicate your knowledge of the subject matter and your ability to apply it to a specific question or situation. Your essay should be well developed and organized, and clearly focused on the question presented. You should reach the right conclusion for the right reasons. Proper grammar, punctuation, spelling, and style are all important to an effective essay. If your writing skills are deficient, then you will have a difficult time with essay questions and you owe it to yourself to get help. An inability to express yourself effectively in writing is a serious impediment to advancement in the business world.

IRAC stands for **Issue Rule Analysis Conclusion**. IRAC methodology is a powerful way to organize and write essays requiring legal reasoning and/or critical thinking. In fact, it is adaptable to any subject matter. In law, IRAC is applied to a set of facts to determine what the proper legal outcome should be.

ISSUE: The issue is the legal question presented. What legal question is raised by the facts? There can be more than one. Frequently, you will be given the issue. If you are given the issue, write about it. Don't change the issue and write about something else. If the question simply says "Discuss," then you must determine and properly state the issue.

RULES: Once you know the legal issue, then you can determine the relevant law necessary to answer the legal question presented. In this section you should accurately and fully state the law that you have determined to be relevant. Relevant rules are those that relate to the development of a party's position or another party's defense or alternative position. Select the rules that support the outcome or conclusion you reach. In an open book or a take-home exam there is no excuse for incorrectly stating the law. If you identify a rule as relevant, I expect you to apply it in the analysis.

ANALYSIS: The analysis is the most important part of your essay. It demonstrates your ability to apply the law to the facts to answer the legal issue presented and to support the conclusion you reach. Analysis is interpretation of facts and law together to determine the objective truth justifying the conclusion. Proper analysis is well organized to logically develop the arguments for and against the issue, and explains why one position is ultimately stronger than another is. Competent analysis accounts for all the given facts, and leaves nothing out in the support of

the conclusion in the same manner that a proof in geometry must include all the steps to be valid. Analysis sets out your reasoning and explains using facts and law together why your conclusion is correct. You will use connecting words like "since" and "because" in your analysis. Omissions in the development of the argument or assumptions in place of facts are usually fatal to your proof. Thus, analysis is your proof that you are right, based on using the proper law and applying it fully and correctly to the evidence (facts and your interpretation of them).

CONCLUSION: The final part of your answer is the conclusion. The conclusion answers the legal question presented in the issue and is your expression of the right or just result or outcome. It should rarely be more than a sentence or two.

Here is a different approach to understanding the IRAC methodology:

When you are reading a case or analyzing a case problem, ask these questions:

1. Who is the Plaintiff/ Defendant(s)?
2. What is/are the claim(s), cause(s) of action, issue(s)?
3. What are the elements of each claim or cause of action?
4. What are the defenses to each claim or cause of action?
5. Is each element proved? Use the facts/evidence to support or negate and explain why.
6. Is each defense proved? Use the facts/evidence to support or negate and explain why.
7. What is the result/conclusion?

If you think about it, question 2 sets the issue, questions 3 and 4 set the rules, questions 5 and 6 set the analysis, and question 6 provides the conclusion. Poof! Instant organization of your answer.

Unfortunately, many people suffer from a lack of writing skills. You cannot become a good writer unless you practice. Here is an easy and fun practice exercise:

Here is a short question followed by a model answer written by a student to demonstrate the IRAC methodology. Following the Model Answer is a Writing Diagnostic for an IRAC answer.

QUESTION

FACTS: Seller owns a forty-acre tract of land. Buyer, after passing the land one says, sends a letter to Seller which states: "I like your property very much. Would you like to accept $60,000 for it?" Seller immediately writes back: "Yes, I would. I accept your offer of $60,000 for my forty acres." Has a contract been formed?

MODEL ANSWER

ISSUE: Did Buyer make an offer to Seller that Seller accepted to create a contract?

RULES OF LAW: An offer is a promise or commitment to do or refrain from doing some specified thing in the future. An offer must be reasonably definite or certain so that the parties can ascertain the terms and must communicate a serious intent by the offer or to be bound by its terms. Statements of interest or inquiry are not offers but are invitations to negotiate. If there isn't an offer, there can be no acceptance.

ANALYSIS: Buyer has not made an offer to purchase Seller's 40 acres for $60,000 because he has not expressed a serious intent to be bound by its terms, even though the terms are reasonably certain. The terms are reasonably certain because a definite price is stated and, although the property could be described more specifically, such as location, the parties can probably determine that they are referring to the same parcel of land.

By stating that he liked the property very much, Buyer has merely expressed his interest in the land. Buyer then inquires whether Seller would take $60,000 for the land. This question is in the nature of a probe or trial balloon and may provide a basis for further negotiations, but it is not an offer.

Neither statement by Buyer in his letter indicates a serious intent to be bound. The first statement is irrelevant to intent. A person can like many things without intending or desiring to acquire them. The second statement is tentative, rather than definite. It invites further discussion rather than express an absolute intent to be bound.

Since no offer has been made by Buyer, Seller cannot "accept" and create a binding agreement. All Seller has done is communicate his willingness to sell and reveal to the Buyer that a contract might very well

be concluded at a $60,000 purchase price. Seller's letter provides a basis for further discussion to form a binding contract.

CONCLUSION: Since there has been neither an offer nor an acceptance, no contract has been formed between Buyer and Seller for the purchase of 40 acres of land.

2.4. Communication

Good communication is essential to good business. A professional must write and speak well.

A design professional must speak and write with precision, but must also learn how to recognize and respond to ambiguity. Ambiguity is also sometimes a useful tool in negotiation and even occasionally in contracts.

The primary business tool of any business is the contract. A significant portion of the course is about contract law and contract provisions relevant to the business of the design professional. The design professional must know and understand the language commonly used (and often abused) in contracts.

A design professional must be a skilled negotiator and mediator.

A design professional must be an expert listener.

Few people are born with the proverbial silver spoon in mouth. Public speaking is an acquired skill. It requires practice.

People must learn how to write well. Writing requires practice.

Are you willing to do what it takes to become a good communicator? Never stop working on your communication skills. Always PRACTICE.

2.5. Constructing a Paragraph

Unfortunately, many people suffer from a lack of writing skills. You cannot become a good writer unless you practice.

Are you a good writer? Show me! Here is an easy and fun practice exercise:

ASSIGNMENT

Write a six or seven sentence paragraph recommending a favorite CD. You <u>must</u> follow these directions:

1. The first sentence introduces the name of the album and the artist.
2. The second sentence describes the kind or type of music on the album and can include something about the artist. However, you may need a separate sentence to describe both the music and the artist.
3. The next three sentences set forth the reasons why you like the album and are recommending it. Be strong and specific in your descriptions so that the reader can gain some understanding of what you are hearing and why the album is so good. For example, you could describe different tracks on the album.
4. The last sentence brings your recommendation to a conclusion. It should provide some final insight in support of your recommendation rather than something lame like "So in conclusion you should buy this CD."

Not a music fan? No excuse. Adapt the assignment to recommend a favorite movie, book, or video game instead. If you can't think of something fun to write about and recommend with enthusiasm to someone else, you desperately need a hobby.

Use the writing diagnostic attached at the end of this article to evaluate your paragraph. You may wish to ask a friend to do this for you. The only way to improve your writing is through practice. You must write, have your writing critiqued, correct your writing and have it critiqued again.

If you cannot easily do this assignment, you need to get help to improve your writing to do well in this course.

2.6. Paraphrasing

Paraphrasing is an effective way to determine whether you have understood someone or that someone has understood you. Paraphrasing is explaining what you have just read or heard in other words.

Example: "Ignorance of the law is no excuse."

Paraphrase: One is required to know the law and follow it.

Example: "The pen is mightier than the sword."

Possible Paraphrase: Ideas are stronger than force.

How would you paraphrase it? Put your understanding here:

Example: "Give me liberty or give me death."
Possible Paraphrase: Freedom is worth dying for. Put your paraphrase here:

Later in this course, you will be asked to paraphrase detailed contract clauses as part of your learning experience.

2.7. Critical Thinking[1]

What does it mean to be a professional? What does it mean to you to be a professional? That's not the same question, is it?

What does it mean to think critically? Do you think you know how to do it? Do you actually do it? That's not the same question, is it?

Engineers concerned with good thinking must be as rigorous in their approach as lawyers are in their approach. A critical thinker in any discipline must possess certain intellectual traits and regularly and routinely apply intellectual standards to elements of thought.

The standards are clarity, accuracy, precision, relevance, depth, breadth, logic, significance, fairness, and ethics. What do each of these terms mean?

The elements are purposes, questions, points of view, information, inferences, concepts, implications, and assumptions. What do each of these terms mean?

The required traits are intellectual humility, intellectual autonomy, intellectual integrity, intellectual courage, intellectual perseverance, intellectual empathy, confidence in reason, and fair-mindedness.

Egocentric thinking and action are to be avoided by the true professional.

You will be asked to make use of the above standards, elements and traits in this course and throughout your professional and personal life.

---

1 Adapted from The Thinker's Guide to Engineering Reasoning, Richard Paul, Robert Niewoehner, and Linda Elder, Foundation for Critical Thinking, 2006

WRITING DIAGNOSTIC

TEST NUMBER_____

GRADE_____

FACTORS: Yes/No

Ratings: Superior, Excellent, Average, Fair, Poor

- Y/N Followed directions for the question
- Y/N Followed IRAC Format
- Y/N Stated the Issue if not given in question
- Y/N The Issue is well and correctly stated
- Y/N Relevant rules are correctly stated
- Y/N Relevant rules are fully stated
- Y/N Facts are fully considered in analysis
- Y/N Facts are correctly stated in analysis
- Y/N Facts and law are used together to support conclusion
- Y/N Rules are applied properly to facts
- Y/N All elements of relevant cause of action are discussed
- Y/N All possible defenses are discussed
- Y/N The analysis is logically organized and presented
- Y/N The analysis is well-reasoned
- Y/N Required or permissible inferences are made
- Y/N The conclusion is correctly stated
- Y/N Structure of the paper is clear
- Y/N Each paragraph is coherent
- Y/N Transitions from one idea/element to next are logical
- Y/N Sentence structure is correct
- Y/N Sentences are well written and varied (not awkward)
- Y/N Word choices are correct
- Y/N Punctuation errors are few
- Y/N Spelling errors are few
- Y/N Grammar errors are few

Result: The conclusion is correct/incorrect and/but
- All relevant material is covered fully and correctly
- Substantially all relevant material is covered correctly
- Some relevant material is omitted and/or there are weaknesses or errors in the analysis
- Significant relevant material is omitted and/or there are major weaknesses or errors in the analysis
- The analysis is general and conclusionary
- The analysis is superficial
- Is unsupported by the analysis
- The analysis is so wrong, incomplete or disorganized as to be nonexistent

Comments:

3: Tort Law

By John W. Tulac

3.1. Introduction

A tort is a civil wrong for which the law provides a remedy.

Everyone is free to do as one wants, provided one does not interfere with the equal freedom of others. Tort law protects people and property by imposing reasonable limitations on the otherwise unbridled freedom to act.

We may have the power to do something, but not the right to do it under a given set of circumstances. For example, I can possess a gun and have the power to use it, but I do not have the right to point it at you and demand your laptop computer.

This article is at most an outline of the most important aspects of tort law. You are required to use the Internet to learn more about torts at web sites such as findlaw.com, wikipedia.com and nolo.com. These are examples. Run a search for different terms. There is a wealth of free information available.

3.2. Classification of Torts

Torts are classified as intentional, negligent, and strict. Torts can be further classified as acts harmful to persons, acts harmful to property, or acts harmful to both.

3.2.1. Intentional Torts

An intentional tort is one that is purposely done. The actor's intent does not actually have to be to cause actual harm. It is sufficient that the person doing the act knew or *should have known* that the act was likely to cause harm.

Examples of intentional torts:
- Assault
- Battery
- Conversion
- Defamation
- Fraud
- Intentional Infliction of Emotional Distress
- Invasion of Privacy
- Nuisance
- Sexual Harassment
- Trespass to Property

"Intentional torts" require that the plaintiff prove the element of intent.

There are two types of rules to define intent:
1. Doing an act for the purpose of causing the consequences
2. Doing an act that creates a substantial probability (or substantial certainty) of causing the damages.
   a) Presumption: a person is presumed to have intended the natural and probable consequences of his or her conduct
   b) Transferred intent
   c) Distinguish proof of motive from proof of intent

3.2.2. Negligence

A negligent act is one that is done carelessly, recklessly, thoughtlessly, or foolishly and results in harm. The actor did not intend to cause harm, but should have known better than to act that way.

We do a lot of stupid things, but most of the time we don't cause any harm to others or to property. You've heard the phrase "No harm; no foul." Negligence is like that.

3.2.3. Strict Liability

Under certain circumstances, liability can be absolute if a harm or loss results, regardless of intent or the amount of caution or control we used. Certain activities are ultra-hazardous, but nevertheless have beneficial effects. We permit such activities but hold the operator strictly liable for any resulting harm or loss. This is known as strict liability. Product liability is an example of strict liability.

3.3. Legal Duty and the Standard of Care

Tort law is based on the concept of legal duty. Everyone must follow the legal duty to act around others in a reasonable way as a reasonable person would under the circumstances. Legal duty is all about knowing how to act (or sometimes to refrain from acting) under the circumstances.

The standard of care, which defines a specific legal duty for specific circumstances, is based on reasonableness, not perfection. Ask: What would an ordinary person of ordinary means and ability do under the same or similar circumstances? The standard of care is the objective description of the legal duty. It tells us what to do. It sometimes tells us how to do it.

Legal duty as measured by the standard of care moves from the general to the specific.

Example: A driver of a motor vehicle must always operate and maintain the vehicle in such a way as to avoid harming others or property. This is the general duty of care for drivers of motor vehicles.

A reasonable person would never drive faster than it is safe to drive for the circumstances and within any speed limit designated for the area. This is a specific standard of care for how to drive a motor vehicle.

A reasonable person will not exceed 25 miles per hour in a residential area and will slow to 15 miles per hour in a school zone when children are present. This is an even more specific standard of care for more specific circumstances.

3.4. The Professional Standard of Care

A professional is held to a higher standard of care for activities directly connected to the professional's specialized expertise. The higher standard applies only to what the professional does as a professional in the capacity as a professional. The higher standard of care does not apply to ordinary activities of the professional. For example, the professional is held to the ordinary standard of care for operating a vehicle even if it is being used "on the job."

Ask: What would a professional of ordinary means and ability within the area of professional expertise do under the same or similar circumstances in the same or substantially same geographic area?

The conduct of a surgeon in a rural area with a properly equipped hospital cannot be directly compared with the conduct of a surgeon at UCLA practicing surgery with state of the art equipment and facilities. This would not be fair. Like must be compared to like.

However, if a professional claims to be more than he or she is, that professional can be held to the standard of the higher level of expertise that was claimed. If a professional licensed in one state acts in another state (or, more likely, under the licensed authority of a professional in another state), that out-of-state professional can be held to a higher standard of care than if the same work was done in the home state.

3.4.1. California Jury Instructions

The follow jury instruction (BAJI 6.37, 6.37.1 & 6.37.2) is that given in California Superior Courts to describe the professional standard of care.

BAJI 6.37

In performing professional services for a client a [specific professional] has the duty to have that degree of learning and skill ordinarily possessed by reputable [specific design professional] practicing in the same or a similar locality and under similar circumstances.

It is a further duty to use the care and skill ordinarily used in like cases by reputable members of the same profession practicing in the same or a similar locality under similar circumstances, and to use reasonable diligence and best judgment in the exercise of professional skill and in the application of learning, in an effort to accomplish the purpose for which the professional was employed.

A failure to fulfill any such duty is negligence.

This jury instruction has been problematic for a couple of reasons. First, the jury must be given evidence of what is a "reputable" member of the specific profession.

Second, the expression "best judgment" implies the highest possible competence or decision-making when, in fact, the law does not require perfection. The exercise of best judgment of a junior engineer may be perfectly competent and within the standard of care, even though a more experienced engineer or one of the objectively best engineers in the country might exercise best judgment in a different way. Therefore, expert testimony is always required to determine whether a design professional has breached a duty of care.

The final sentence of the jury instruction is technically incorrect. The failure to fulfill a duty of care is a breach of the duty of care, but is not per se negligence. Negligence requires proof of causation as well.

BAJI 6.37.1

It is the duty of a [specific professional] who holds himself or herself out as a specialist in a particular field of [specify], to have the knowledge and skill ordinarily possessed, and to use the care and skill ordinarily used by reputable specialists practicing in the same field and in the same or a similar locality and under similar circumstances.

This instruction is often given when the case involved a specific specialty or subfield within which a professional is practicing.

BAJI 6.37.2

A [specific professional] is not necessarily negligent because he or she errs in judgment or because his or her efforts prove unsuccessful. However, a [specific professional] is negligent if the error in judgment or lack of success is due to a failure to perform any of the duties as defined in these instructions.

A professional is not a guarantor of results and a professional's work is not judged by a standard of perfection.

The most difficult task sometimes is to go from a general description of the standard of care to a specific description of the standard of care for specific circumstances. For example, our general legal duty when we drive a motor vehicle is to always drive it in a manner that is safe under the circumstances and to never go beyond any stated maximum like a posted speed limit. For the specific circumstances, such as a school zone or a blizzard, we further describe the specific legal duty. The DMV conveniently gathers such rules for us in a nice little pamphlet. Thus, when we park a car on the hill we know that we have the specific legal duty to put the car in park, set the emergency brake and turn the wheels in the appropriate direction.

3.4.2. Review Exercise for the Professional Standard of Carey Smith

So now, let's see how well you can determine the professional standard of care. Assume you are a licensed civil engineer based in Los Angeles, California:

1. Define the general professional legal duty of care you must follow.
2. You are asked to provide design services for a new bridge over the Los Angeles River using new lightweight composite materials that have not previously been used in bridge construction. Define the specific standard of care you must follow.

3. You are asked to design a conventional arch-type bridge over the River Kwai in Thailand. Assuming Thailand has laws similar to the United States, define the specific standard of care you must follow.
4. You park on the street to visit a client that has an office near the top of a steep hill. Define the specific standard of care you must follow.
5. You contract with the owner of the property to provide consulting services reviewing the plans and specifications of a licensed architect and a licensed civil engineer for a building to be constructed in Corona, California. Define the specific standard of care you must follow.

You should need no more than one sentence to answer each of these subparts.

3.5. General Structure and Analysis of Tort Liability[1]

3.5.1. Elements of the Plaintiff's Cause of Action

Each tort can be broken down into basically four "sub-parts" or "sections":

1. Act or conduct requirements:
   a) These elements (there may be more than one "act" or "conduct" element) define what the plaintiff must show the defendant to have done, not done or what the defendant must have known at the time.
   b) They define the nature and quality of the defendant's conduct.

   Example
   Assault: A physical act that creates in another person a reasonable apprehension or fear of an immediate harmful or offensive touching.
   Note: This definition of the "act" or "conduct" requirements can actually be viewed as a series of required elements:
       i) A physical act
       ii) Creating reasonable apprehension or fear in another person

    iii) Of an immediate harmful or offensive touching
   Note: When analyzing the conduct of an individual, be sure that their conduct conforms to all of the conduct requirements, including "sub-elements."

2. Tort categorization
   a) Is it a tort that requires the plaintiff to prove intent?
   b) OR is it a tort that could be committed without fault, negligently or intentionally?
   c) All torts other than the tort of negligence would fall into one of the two categories above

   Examples:
Assault:	Intent required
Trespass:	Intentional
	Negligent
	Without fault

3. Causation:
   a) All torts require the plaintiff to prove that the defendant's conduct was the actual and proximate cause of the plaintiff's injury.
   b) Both parts of causation must be established or proof of causation fails

4. Injury and plaintiff's remedy:
   a) Damages
   b) Injunctive relief
   c) There must always be an injury, and the usual remedy is damages.

3.5.2. Affirmative Defenses

Next, you should consider all possible affirmative defenses that relate to the tort.

For example, the doctrine of consent and the doctrine of self-defense are affirmative defenses to all intentional torts.

To some degree, the affirmative defenses vary depending on the tort the defendant is alleged to have committed.

3.6. The Remedy of Damages

Remember "no harm; no foul" as a general proposition.

Critical Rules

1. Actual or compensatory damages:
   a) Compensation for injury that would naturally flow from the defendant's conduct
   b) Personal injury to the plaintiff, the plaintiff's property, reputation, privacy, business interest, pain and suffering, emotional distress, etc.

2. Punitive damages:
   a) Generally requires proof of a form of intent—more than ordinary negligence
   b) Purpose of punitive damages
   c) Economic significance: insurance and bankruptcy

3. Injunctive Relief
   a) Injunctions are an "Equitable" Remedy
   b) Prerequisites of Equitable Remedies
   c) Negative vs. Affirmative Injunctions
   d) Types of Injunctions

3.7. The Element of Causation

Critical Rules

1. Actual cause
   a) Was the defendant the actual cause in fact of the plaintiff's damages?
   b) Two tests:
      i) *But For Rule:* But for the defendant's conduct, would the plaintiff's injuries have occurred?
      ii) *The Substantial Factor Test:* Was the defendant's conduct a substantial factor in bringing about the plaintiff's injury
         I *Only Factor:* Actual cause need not be the "sole contributing factor"
         II *Most Substantial Factor:* Actual cause need not be the "most" substantial factor
         III *Comparative Liability:* Can result where damage results from more than one substantial factor

2. Proximate cause of "legal cause":
   a) To what extent should the defendant be liable for the damages for which he was the actual cause? This is a "social policy issue."
   b) Foreseeability of the injury: Was injury to the plaintiff "foreseeable" at the time of the defendant's conduct?
   c) Intervening forces: Sometimes intervening forces (forces that occur after the defendant's original conduct) increase damages. Will the defendant be liable for these increased damages? Yes, unless the intervening forces are "supervening forces."
   d) Supervening forces: What intervening forces are supervening forces?

Example: There is no liability for a third party's intervening intentional tort that was not foreseeable. This "supervening force" is said to "break the chain of causation."

3.8. Negligence

1. Act:
   An action or failure to act (when under a duty to act) that raises a risk of harm to another.

2. Duty:
   a) To foreseeable individuals
   b) To unforeseeable individuals where
      • No one was foreseeable
      • Someone was foreseeable, but not the plaintiff

3. Breach of standard of care:
   a) The general duty rule:
      • One has a duty to act as a reasonable, prudent person would act under the same or similar circumstances.
   b) Special duty rules for special circumstances
      i) Negligence Per Se:
         • Breach of a statute or ordinance intended to protect a class of individuals that would include the plaintiff.
      ii) Res Ipsa Loquitur:
      Facts speak for themselves. Applies when:

- The event creating the damage is one that ordinarily does not occur in the absence of negligence; AND
- All the instrumentalities of causation include the defendants.

iii) Other Special Duty Rules Exist

Examples: Duty of owners and occupiers of land to trespassers, licensees and invitees.

4. Causation:
- See prior discussion of actual and proximate cause. Causation is an element of every tort.

5. Injury—Damages:
(See prior discussion of "actual or consequential damages." A plaintiff **cannot** obtain punitive damages for negligence.)

NOTE: If the Plaintiff has established each element of the cause of action for negligence, the burden of proof has been met and a prima facie case for the defendant's negligence has been made. The burden of proof then shifts to the defendant to show why liability should not be imposed.

Defendant can raise one or more affirmative defenses to show either that Plaintiff is in whole or part responsible for her own harm or loss or that someone other than Defendant is responsible in whole or part for the plaintiff's harm or loss or both Plaintiff and one or more other parties are in whole or part responsible.

Affirmative Defenses:
6. Assumption of the Risk:
  a) Knowledge of the risk
  b) Voluntary assumption of the risk

7. Comparative Fault (Contributory Negligence):
  a) Failure to exercise the degree of care and caution that a reasonable, prudent person would exercise on behalf of himself or herself
  b) The doctrine of comparative negligence: By showing the comparative fault of another (often the plaintiff), the defendant can effectively lower his or her liability.

3.9. Professional Negligence

Professional negligence is established in the same way as ordinary negligence, but under the measure of the professional standard of care.

3.10. Fraudulent Misrepresentation

1. Act: Misrepresentation of facts or conditions
  a) With knowledge that they are false; OR
  b) With reckless disregard for the truth; OR
  c) Failure to disclose when under a duty to disclose.

2. Intent: *Scienter* is intent to induce another to rely on the misrepresentation.
3. Justifiable reliance by the deceived party:
4. Causation: same as above
5. Injury—Damages: Actual and punitive damages
6. Affirmative defenses:
  a) Consent
  b) Puffing

3.11. Innocent Misrepresentation or Negligent Misrepresentation

1. Act: Misrepresentation of facts or conditions
2. Intent: This tort can be committed without fault or negligently. If committed intentionally with justifiable reliance, it becomes a tort of fraud for which punitive damages can be awarded.
3. Actual reliance by the plaintiff
4. Causation
5. Injury—Damages: Actual or consequential damages only.
6. Affirmative defenses:
  a) Consent
  b) Puffing

3.12. Strict Statutory Product Liability

1. Act: The product must be in a defective condition at the time of sale.
2. The defendant normally engaged in business of selling such products.
3. The product must be unreasonably dangerous* to the user and consumer because of its defective condition.

4. *This element is not required in California and some other states.

5. The plaintiff must incur physical harm to self or property from the use or consumption of the product.

6. No substantial change of the goods from the time the product was sold to the time of injury.

7. Causation: same analysis as above

8. Injury—Damages:

  a) Actual or compensatory damages only (in the absence of an intentional tort theory).

  b) Affirmative defenses and limitations on recovery:

    i) *Assumption of Risk:* (See previous discussion)

    ii) *Misuse of the Product:* Similar to assumption of the risk—limited by courts if misuse is reasonably foreseeable

    iii) *Comparative Fault:* E.g., plaintiff who fails to take precautions against a known defect may have recovery reduced

    iv) *Statutes of Repose:* The limited time within which a plaintiff can file a product liability suit begins to run on an earlier date than the statute of limitations

    v) *Bystander Liability:* May require "reasonable foreseeability"

3.13. Tort Reform

1. Limits on non-economic damages—such as pain and suffering
2. Limitations on attorney contingency fees
3. Cap on punitive damages

Are such limitations just?

NOTES

[1] Sections 3.5 through 3.13 based on *Class Notes for FRL 201: Legal Environment of Business Transactions,* Richard Bergstrom (2002)

4: Risk

By William A. Kitch and John W. Tulac

4.1. *What is risk?*

An engineer might define risk as a quantitative value using a formula. However, a lawyer would view this approach as too narrow. Risk is also qualitative in nature. Risk springs not only from averaged or calculated results, but from deviations from the average or norm, from the outliers at the far reaches of the tails, and from outcomes never even imagined. Peter Bernstein's *Against the Gods, The Remarkable Story of Risk* is an excellent reference for engineers and businesspeople alike.

The purpose of this article is to develop a framework to use to indentify, evaluate and manage risks. Those who can manage risk will be able to seize opportunities from those who depend upon luck or chance.

4.2. *Components of Risk*

There are many definitions for risk. However, most definitions contain the following components:

- Danger: cost, loss, or damage
- Uncertainty: probability, possibility, statistical chance
- Opportunity: potential benefits

We will develop our formal definition of risk later using these components.

4.3. *Process of Risk Management*

If we are going to successfully manage risks, we need a process we can follow to ensure our risk management is effective. We will be using the following process:

1. Identification
2. Evaluation: qualitative or quantitative
3. Allocation: determining who has the best ability to manage a specific risk
4. Mitigation: techniques to determine, minimize and prevent harm or loss including:
   - Insurance: professional, business, legal
   - Preventive law
   - Compliance planning

We will now discuss the steps in the process.

4.4. *Identification*

Identification is the most critical step in the risk management process. If we fail to identify a key risk, the rest of our planning process may be meaningless. We can spend all the time we want evaluating a risk, but if we fail to evaluate the important risks, our work will have little value.

It will be impossible to identify risk if you do not know what your goals and objectives are. Therefore before trying to identify your risk, you must first clearly articulate your goals and objectives.

Once you have finished defining your goals and objectives, list the key steps or processes that will be needed to achieve your goals. Then list all the key players in the processes. Now you're ready to identify your risks. Simply make a list of all the actions or things that could happen which would prevent you from achieving your goals. This is a brainstorming time. Don't toss out a risk because it sounds silly or you don't think it will really happen. Just identify the potential negative things and write them down. We will evaluate the risks in the next step.

Risk analysis bridges tort law and contract law. Tort law defines responsibilities and liabilities arising out of risk and imposes duties. Contract law allocates responsibilities and liabilities regarding risk.

4.5. Evaluation

Once we have a comprehensive list of risks we can start the evaluation process. It is important that we use a systematic and consistent method to evaluate the risks. We can use subjective measurements to evaluate the risks so long as we apply our measurements in a systematic manner.

4.5.1. Measuring Risk

If we are going to make logical decisions about how to manage risk, we must first be able to measure risk. Mathematically, risk is defined as

$$R = P_f \times C_f \qquad (4.1)$$

where

R is the numerical risk
P_f is the probability of failure and
C_f is the cost of failure, in dollars

People often confuse risk with the probability of failure or with the cost of failure. According to Equation 4.1, risk, R, is a combination of both the probability and cost of failure. This is a very important concept. There are some things that have a very high cost (like death) but which are very unlikely. An example might be being caught in a bombing on the Cal Poly Pomona campus. The cost of such an event could be your life, but the probability of this event is very low, therefore the risk is relatively low. Another type of event that can have a very low risk is something that is very likely, but which carries with it little cost. For example if you park on campus without purchasing a pass, the probability of getting a ticket is fairly high. However the cost of the ticket is relatively low (at least compared with many other costs associated with owning and operating an automobile). You

may still wish to avoid this risk by purchasing a pass and that's the whole point of risk assessment. If you can quantify your risk in terms of a likely cost, then you can determine whether or not it is worthwhile to pay for actions that can reduce your risk.

In some business it is possible to make very good measures of the actual probability and cost of events and use these measures to compute accurate values for the risk of events. Insurance companies do this all the time. That is the only way that can set premiums and make money. However, it is often impossible or too costly to get accurate measurements of the probabilities and/or costs of events. When you can't quantitatively measure risk, a qualitative measurement of risk is usually possible. While qualitative risk measurement is less accurate than quantitative risk measurement, it can still be a valuable tool in managing risk

4.5.2. Qualitative Risk Measurement: Risk Matrix

One simple and effective method of qualitative risk measurement is to build a risk matrix. Let's describe the process using an example. Suppose we want to evaluate the risk of parking on campus. The first step would be to define our goals and objectives. For our example we will assume our goals are to find a safe parking place that allows us to get to class on time and stay within our budget.

Next we need to list every event that would cause failure or be a problem when attempting to park on campus. That list might look something like this:

1. Can't find a place in time and are late to class
2. Can't find a place in time and are late to an exam
3. Car is damaged in parking lot
4. Car is stolen from parking lot
5. Receive a ticket for illegal parking
6. Forget where you parked and are late getting home

The next step is to make up a qualitative scale for the probability of each event happening and a qualitative scale for the cost of an event should it happen. Let's use the following scales

Probability:
a) Unlikely: Probably happens less than once a year

b) Likely: Probably happens once a month

c) Very Likely: Probably happens once a week or more often

Cost:

a) Low: Painless, wouldn't even count the cost

b) Moderate: Ouch, that's costly, but won't break the bank

c) High: Oh no, I can't afford that

Now for each of our events we rate its probability and cost.

Event	Probability	Cost
1) Late to class	Very Likely	Low
2) Late to exam	Likely	Moderate
3) Car damage	Likely	Moderate
4) Car stolen	Unlikely	High
5) Get ticket	Likely	Moderate
6) Late getting home	Unlikely	Low

We now evaluate the risk by plotting the events on a risk matrix. The risk matrix is simply a two dimensional figure with the probability listed on one axis and cost on the other axis. Figure 4.1 depicts a risk matrix for our example.

In the risk matrix, events that lie in the upper right hand portion of the matrix are the riskiest events. Those that lie in the lower left hand portion are the least risky. In this example we have arbitrarily divided the risk into the five levels shown. Using our example we would categorize our risks as shown in Table 4.1.

Using this analysis method we discover that our highest risk is being late to an exam. If we simply evaluated the risk based on cost, we might have said that getting our car stolen was the biggest risk. But that's not the case; while it has the biggest cost it also has a low probability. Also note that being late for an exam and class both have high probability, but one has a larger cost and is therefore more risky.

When evaluating risk it is important to include both cost and probability. Risk is a combination of the two. It is also important to use a systematic method to evaluate risk. Finally, we don't need exact numbers on

cost and risk to perform an evaluation. Even a simple qualitative evaluation like the risk matrix can help us to identify and manage risks.

4.5.3. Allocation

Risk allocation is the process of identifying which person or party has the best ability to manage each risk and who benefits if a risk is avoided. We then assign risk based on a party's ability to manage and benefit from the management of that risk.

As we look at each risk we will find that they generally fall into one of the following cases.

a) One party has ability to both manage the risk and benefit from managing it

b) One party has the ability to manage the risk but another party is the main beneficiary

c) The management and benefits associated with a risk are shared among two or more parties

d) No one has control over the risk

Table 4.1: Tabulated risks from example

Risk Level	Events
Very High	None
High	1) Late to class, 2) Late to exam
Moderate	3) Car damage, 4) Car stolen
Low	5) Get ticket
Very Low	6) Late getting home

It should be obvious in case-a that the risk should be assigned to the party who can manage and benefit from the risk. In case-b we should try and change processes to align the management and benefit. Many risks fall into case-c and they must be spread appropriately among the responsible parties. Case-d is a situation where our only ability to manage the risk may be to pay someone to take the risk for us by buying insurance.

In our example, the risks associated with Events 1, 2, 5, and 6 clearly belong to the person parking the car, since that person has the ability to manage these risks and will benefit if they are managed. The parker has some control over Events 3 and 4, but much of the risk of these events is beyond the control of the parker.

4.5.4. Mitigation

Now that we have identified, evaluated and assigned our risks it is time to develop mitigations which will allow us to control and manage our risks. We will use our example to illustrate this process.

The most important risk for us to manage is our highest risk: missing an exam, Event 2. There are two obvious mitigations for this risk.
1. Allow extra travel time to campus on exam days
2. Buy a parking pass so we can park close to campus

Notice that these two mitigations will also manage the risk of being late to a regular class, Event 1, and the risk of getting a ticket, Event 5.

We have two other moderate risks that we should mitigate: Event 3–Car damage, and Event 4–Stolen car. We have some ability to manage these risks by choosing safer parking spots, locking our car, and not leaving valuables in our car, but much of the risk is not under our control. This is a case where buying insurance might be a good way to mitigate our risk.

Your mitigation doesn't end when you devised your mitigations. If you don't implement your mitigations they won't do you any good. It is critical that you establish processes and procedures to ensure your mitigations are properly implemented. For example, our mitigation to leave early for campus on exam days won't do any good if we forget or don't know when we have an exam. Our implementation plan for this mitigation might be to put all our exams in our iPhone at the beginning of each quarter with an alarm the day before the exam. This will help ensure we leave in time to park and get to our exams.

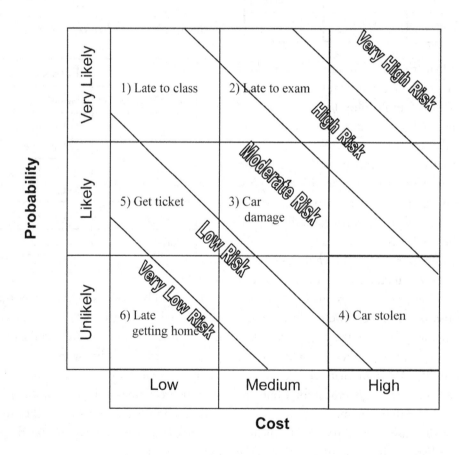

Figure 4.1: Risk Matrix

4.6. Risk Management Tools for the Design Professional

The generic tools such as the risk matrix are valuable for design professionals and should be part of your risk management tool kit. However, there are some other tools you should be aware of.

Perhaps the most valuable tool for risk management is the contract. A formal contract lets you handle risk professionally. It can be used to allocate risks by informing the parties of their responsibilities. It should identify things outside the control of the parties, particularly unpreventable or uninsurable risks. One of the greatest sources of risk is misunderstandings between parties. A well written contract can significantly reduce the probability of misunderstandings, thereby reducing risk. Finally, a contract can be used to establish mitigations, particularly complicated mitigations that require the cooperation and coordination of several parties.

One of the most powerful risk management tools available to anyone is to just say "No." Walking away from a bad deal is the ultimate risk eliminator.

4.7. Compliance Planning

Regulatory Compliance carries significant risks to today's design professional. There are a myriad of local, state, and federal laws that require compliance. These regulations cover many areas including: environmental compliance, employment law, fiscal accounting and reporting. In some cases failure to comply with regulations can bankrupt a company or put a company officer in jail.

Regulatory compliance programs are designed to
- Prevent violations
- Detect violations early
- Respond quickly to any violation
- Report compliance to regulators

Steps to establishing a corporate compliance program
1. Assess activities in areas subject to regulation
2. Obtain the commitment internally from all involved parties
3. Create policies and programs for compliance
4. Implement the program

a) Education and training
b) Controls, procedures, and recordkeeping
c) Review and revise regularly
d) Audit (internal, external)
e) Reporting and disclosure system
f) Revise and update (circles back to step 1)

The adage "an ounce of prevention is worth a pound of cure" applies to regulatory compliance. A good compliance plan can mitigate many risks.

4.8. Summary

Risks are unavoidable. They are part of our everyday life. Risks present opportunities to those who can manage them well. Rather than avoid all risks our objective should be to manage risks for our benefit.

Risk is not just the probability of failure. It is the combination of the probability of failure and the cost of that failure. In order to manage risks we must first identify the risks, evaluate each risk, and then allocate the risk to the appropriate party. We can then develop a mitigation plan to help us manage risks. The plan must be properly implemented or it is of no value.

Perhaps the most important risk management tool for the design professional is a contract. The contract helps to identify, allocate and mitigate risks.

References

The Risk Management Toolkit, Mitre Corporation, http://www.mitre.org/work/sepo/toolkits/risk/index. html

5: Summary of Contract Law

By John W. Tulac

Studies have shown that the two most important components for an economy to develop are education to create knowledge and a rule of law that protects private contract and private property rights. Property rights, real and personal, tangible and intangible, contribute to the development of wealth by the ability to barter or exchange them. Money is a medium for exchange. The basis by which people are able to get what they want or need is the contract.

A contract creates a binding obligation or commitment between two or more parties for which the law recognizes a duty to perform and provides a remedy in the event of nonperformance of the obligation or commitment.

5.1. Contract Elements

A contract exists when the following four elements are present:

1. Agreement. Agreement, which consists of an offer and an acceptance, forms or makes the contract.
2. Consideration. Consideration is the subject matter of the contract.
3. Contractual Capacity. The parties making the contract must be legally competent to do so.
4. Legality. The subject matter or consideration must be lawful.

In addition, the courts look at two other aspects to determine whether a contract can be enforced. These are affirmative defenses to the enforceability of the contract. They are:

1. Form. Some contracts must be in writing to be enforceable
2. Genuineness of assent. The agreement must be volitional or voluntary, that is entered into by one's own free will that is not wrongfully induced or coerced into agreement.

5.2. Types and Classification of Contracts

Contracts are classified in a variety of ways, including:

Bilateral or unilateral. A bilateral contract is formed by the mutual exchange of promises. Joe promises to mow Jim's lawn and Jim promises to pay Joe $25.00 for mowing the lawn. A unilateral contract is formed by the exchange of a promise and the actual performance by one of the parties. Jim promises to pay Joe $25.00 if Joe mows Jim's lawn. By then mowing the lawn Joe makes the contract and performs it at the same time. Obviously, unilateral contracts can present some risk to the accepting/performing party.

Express or implied. An express contract is one that is written or oral. An implied-in-fact contract arises out of the conduct of the parties. An implied-in-law contract, also called quasi-contract, is not an actual contract at all but rather is a remedy by which a court will enforce some promises upon which the other party is found to have been reasonably entitled to rely. It is smart contracting practice to make express contracts, preferably in writing, even when the law does not require a writing.

Voidable and void contracts. This distinction is the bane of business law students. A voidable contract is a true contract but one or both parties has the right to withdraw from the contract for various legal reasons such as fraud, mutual mistake of fact, undue influence or wrongful coercion. A void contract is not a true contract at all, but rather is a nullity. An illegal contract lacks the element of legal subject matter and therefore is void. A contract to distribute cocaine is a void contract.

Executory and executed contracts. An executory contract has not yet been performed or fully performed. An executed contract has been performed. As is so typical in the English language an executed contract also means a contract that is signed by the parties. Written contracts must be signed, at least by the party to be charged with performance, but obviously good contracting practice requires all parties to sign.

Contracts are subject to interpretation. A contract will be given the plain meaning of its language. Ordinary words are given their ordinary meaning. Specialized words are given their specialized meaning within the relevant context. A contract which is intended to be complete on its face and so states will not require external or parole explanation as to its terms. A contract for the sale of chicken can be fulfilled by any bird that is a chicken, young or old, small or large, roaster or fryer or stew bird. For a particular kind of chicken to be delivered under contract, the parties would need to use an adjective to further describe the subject matter.

There are many other rules for interpretation of contracts terms. Terms are interpreted to preserve the existence of contract, to give it a lawful purpose over an unlawful one, to give more weight to customized or negotiated terms over preprinted or boilerplate terms, to resolve ambiguous terms if possible so that the contract can be enforced, to make use of trade or market customs and practices to explain terms etc.

5.3. Offer and Acceptance
5.3.1. Offer

A contract is made or formed by an offer and an acceptance.

An offer is a promise to do or refrain from doing some specified thing in the future. The promise can be made expressly in words, oral or written, or implied by conduct.

An offer consists of three parts: 1) the offeror must have serious objective intent to bind, 2) the terms of the offer must be reasonably definite and certain, and 3) the offer must be effectively communicated between the parties. Once these three requirements are met, the offer can then be accepted and an agreement can be formed, potentially resulting in a contract, given other elements are met.

Intent is measured by the offeror's objective intent (the party's action and words) and not by their subjective intent (internal assumptions, intentions and beliefs). Offers made in obvious anger, jest or excessive excitement are not considered offers with objective intent because any reasonable person would not presume this to be a serious offer.

Expressions of opinion (such as, "This Honda will probably run like a corvette when we're through with it!"), statements of intent ("I plan to sell my vector dynamics textbook for $80 after this quarter"), preliminary negotiations, agreements to agree, advertisements (these are generally considered as solicitations or invitations to negotiate), and actions are generally not considered contractual offers.

Though the contract's terms may vary, some terms must be either expressly stated in the offer, or capable of being inferred from it. Typically, an offer must identify the parties, the subject matter of the potential contract, the consideration to be paid, the time of payment, and time of delivery or performance. However, the contract may still be enforceable if the parties clearly show intent to form a contract but have left out a particular term, in which case the courts may infer a reasonable term. If the terms are too vague, the court will not implement a term and the contract will not be enforceable. In a contract for the sale of goods, only the quantity term must be expressly stated; all other terms may be implied.

The third and final requirement of an offer is communication of the parties. The offeror must effectively communicate the offer to the offeree. One cannot accept an offer of which she has no knowledge.

There are two means by which an offer may be terminated: via the actions of the parties or operation of law. The offeror may terminate the offer by revocation or the offeree may terminate the offer by rejection.

An offeror generally may revoke the offer at any time prior to acceptance; however, in the event of an option contract, promissory estoppel can render an offer irrevocable for a stated time. With the exception of public offers, revocation is not effective until received by the offeree. Rejection occurs when the offeree demonstrates a clear intent not to be bound by the offer and is not effective until received by the offeror. An offer can also be terminated by a counteroffer by the offeree. It is also possible to make a counteroffer without rejecting the offer, but this is somewhat tricky to do.

An offer can also be terminated by operation of law by one of four ways: lapse of time, destruction, death or incompetence, and illegality. An offer will terminate by either a reasonable period of time or at the end of the stated time. Destruction of the subject matter, illegality of the offer or death / incompetence of either of the parties will also terminate the offer.

5.3.2. Acceptance

In order to complete the agreement, the offeree must accept the offer. An acceptance must be both unequivocal and communicated in order to legally form a contractual agreement.

For an acceptance to be unequivocal, it must follow the mirror image rule: the acceptance must match the terms of the offer. If the offeree makes any material changes to the terms of the offer, it can be deemed a counteroffer, thereby voiding the original offer and by which means the offeree would then become the offeror. However, statements of opinion like "I accept, but I wish it hadn't been so expensive," and inquiries, "I accept, but are you willing to lower the price?" are still considered unequivocal and therefore an acceptance. Whereas a statement like "I accept, but only if you give me a discounted price" is not unequivocal acceptance and would be a rejection and a counteroffer. As long as there are no conditions to acceptance given (as in, "I accept, if there is a discount."), then the acceptance is effective.

Generally, silence cannot act as acceptance, even if the offeror states "By your silence, I conclude you're accepting the offer." However, there are a few instances in which silence may operate as acceptance: when the offeree accepts services rendered by the offeror even though she had the opportunity to reject them and it is reasonably understood there is an expectation of compensation. Such silence would lend itself to the formation of an implied-in-fact contract. The offeree is legally obligated to compensate the offeror for the benefit received. Another means by which silence operates as an acceptance is when the offeree has prior dealings with the offeror.

Acceptance, much the same as an offer, generally requires a communication since most contracts are bilateral. In a bilateral contract, communication of acceptance is necessary. In a unilateral contract, performance serves as acceptance so notification becomes irrelevant unless otherwise requested by the offeror.

The only other acceptable acceptance by silence is with regards to a solicitation: if an offeror makes an offer with definite terms and conditions in response to a solicitation, the offeree must reject the offer or their silence will serve as acceptance.

Finally, the mode and timeliness of acceptance must be considered. The general rule in bilateral contracts is that the acceptance is timely if it occurs before the offer terminates. Typically the timeliness of an acceptance follows the mailbox rule: when the acceptance is deposited in the mail the acceptance becomes valid (if mail is the authorized mode of communication). Authorized means of communication are either expressly or impliedly authorized by circumstantial facts, generally the same manner by which the offer was made or faster (ex: a fax is faster than the mail and is a writing, so a fax is impliedly authorized). As soon as authorized means of communication are used, the contract is formed.

When a mode of communication is not specified, acceptance becomes effective upon receipt of the acceptance and may be made by any reasonable means. Generally when an offeree sends an acceptance by means other than that express or implied, acceptance is effective when received, unless it was timely sent and timely received, then acceptance is effective on dispatch, regardless of means used. However, there are a few exceptions: if the acceptance is not properly dispatched, then it generally is effective only upon being received by the offeror; the offeror can expressly state that acceptance is not effective until receipt by offeror; if the offeree sends both an acceptance and a rejection (perhaps a rejection and then changes their mind) then the first communication received determines the formation of the contract.

5.4. Consideration

Consideration is the subject matter of a contract. Consideration flows from and to each party. Each party gives something. Each party gets something in return.

The primary reason for the enforceability of promises is consideration. In order for consideration to exist, there must be a legal value and a bargained-for exchange. Legal value can include a promise of action that has no prior legal duty, performance one is not otherwise obligated to undertake and forbearance (refraining from performance of an act that is otherwise legal).

However, in order for there to be proper consideration, there must also be a bargained-for exchange. For there to be a bargained-for exchange, the promise from the promissor must be an incentive for the promisee to return the promise. Bargained-for exchange is what distinguishes contracts from gifts.

In other words, it is what each party receives as the benefit of making the contract. Each party gives something; each party gets something in return. Each part of giving and getting is consideration.

If valid consideration exists, courts will rarely examine the adequacy of consideration (the value or amount of consideration given). Generally, however, as long as consideration is legally sufficient (both a bargained-for exchange and of legal value) the courts will not review the consideration unless there is an extreme case in which inadequacy might indicate fraud, duress, or undue influence. In other words, a person is free to make a bad deal and won't be excused from it unless it is wrongfully induced.

Once a party has agreed to terms, it undertakes a "preexisting" duty to perform as agreed despite most changes in circumstances. However, should the parties encounter unforeseen difficulties upon performance of the contract, additional consideration may be allowed to maintain the fair equity of the original contract. For example:

> Joe is hired by NASA as an independent contractor to develop a new telescope. The parties have agreed upon a contract price of $5,000,000 for the development and production of a high-powered telescope. While Joe is building the telescope in area 51, a freak meteor crash-lands on his laboratory, completely destroying his work and maiming Joe in the process. To rebuild the telescope and the laboratory necessary for the process, Joe would have to incur thousands of dollars worth of expenses. This unforeseen (and unforeseeable) difficulty would exempt Joe from the preexisting duty rule, providing a basis for unilateral contract modification or termination.

However, if the economy overheats and inflation results, which increases the costs of materials instead of a rogue meteor crashing into the laboratory, then the courts would assert the preexisting duty rule because the inflation is not an "unforeseen difficulty" but rather a risk customarily assumed in business. However, parties could mutually choose to rescind or cancel the contract.

A promise of consideration made for events that have already occurred lacks bargained-for exchange and thus is unenforceable. For example:

> Beth offers to pay Danielle $1,000 for passing vector statics last quarter. Danielle did not have to perform or forbear anything in order to receive the payment. One cannot bargain for something that has already occurred. Thus, the contract would be unenforceable.

The issue of consideration usually arises when performance by one or more of the parties is uncertain, the settlement of claims, and whether or not promises are enforceable without consideration.

Some promises are illusory. For example:

> Cal Poly Pomona sends out a bulletin stating that, "If a significant portion of the student body attends summer classes, Parking Services will lower the cost of student parking passes by $40." This is an illusory promise given that "a significant portion"

is ambiguous at best and completely up to the discretion of the university. There is no bargained-for exchange and thus, no consideration and no contractual obligation.

A unilateral option-to-cancel clause may also make a promise illusory and the contract unenforceable. A major automobile company once issued new contracts to all its suppliers and made them sign or the suppliers would no longer be used. The contracts essentially said that the auto company would buy whatever it wanted in whatever quantities it wanted whenever it wanted (or not). A supplier successfully challenged the contract as not truly a contract at all.

Consideration can nevertheless be valid even when the exact performance is uncertain, as is common in requirements and output contracts. In brief, a requirements contract is when the seller provides the buyer with the goods of any given type that the buyer "requires." An output contract is a contract in which the buyer agrees to purchase all of the seller's output or product. These types of contracts are typical of oil and energy contracts. For example:

> The Big One Oil Co. builds a new oil refinery station in the middle of Alaska. The station requires coal to operate, so The Big One Oil Co. contracts with Little Timmy Mining to supply the refinery station with all of its coal-fueling requirements for the next 20 years. This would be an example of a requirements contract.

> If Texaco contracts to buy all the refined oil products Big Oil produces at its Alaskan refinery, it has made an output contract.

It is important to understand the nature consideration for a number of reasons, one of which is settlement. Settlement usually includes accord and satisfaction, or a release and covenant not to sue. In the case of debt, the parties can agree to an accord and satisfaction: an agreement in which the debtor offers, and the creditor accepts, payment or performance of something other than what was originally agreed in satisfaction of the claim. . For an accord to occur there must first be a dispute regarding the amount of debt.

From there the parties can create an accord, after which is executed, satisfaction of the debt can take place. If the debt is liquidated, however, an accord and satisfaction cannot take place because the amount of the debt is fixed or has previously been ascertained, as in the case of a bank loan. There is thus no consideration due to the preexisting obligation that would be present in a liquidated debt settlement. And yet, if the debt is unliquidated, or in dispute, then an accord and satisfaction can occur, discharging the debt.

A release, on the other hand, is an agreement that bars any further recovery beyond the agreed upon terms of the release. A release is generally binding if it has consideration given for the release, is made in good faith and is in a signed writing.

> Jesse McDaniel is an engineer engaged by Boss Man Big to design and produce a new fuel injection system for Mr. Big's new line of 2009 high performance super cars. Upon completion of the fuel injection system, the design is implemented into the super cars and production goes into full swing. After a few months, it's discovered that the fuel injection system's design causes the car to misfire during acceleration in 60% of the vehicles. When the problem is traced to McDaniel's faulty design, Big makes a claim against McDaniel. McDaniel then offers Big $100,000 for a release from any further liability. Even if later it is discovered that damages from the faulty fuel injection system design equates to roughly $2.5 million, Big would still be legal barred from suing McDaniel for the recovery of further damages. You would therefore be further ahead to understand the full extent of your damages before signing a release.

In California, a covenant not to sue, unlike a release, does not bar one of the parties from further recovery. A covenant not to sue replaces one contractual obligation with a legal action based on a valid claim.

> Instead of a release, McDaniel agrees to redesign the fuel injection system, correcting

the errors for Big's agreement to not sue for damages. If McDaniel fails to develop a new fuel injection system, then Big can sue for breach of contract.

There are only a few rare cases in which consideration is not required: promises to pay debts barred by the statute of limitations, promises under promissory estoppel and charitable contributions. If a creditor fails to sue for the recovery of a debt within the statute of limitations, the law then bars the creditor from doing so after this period. However, the debtor can agree to pay a past debt regardless of this restriction; in which case no consideration would be necessary. Under the doctrine of promissory estoppel, even though the promise may lack consideration, if the promise is clear and definite, the promise detrimentally and reasonably relied on the promise, the promise is substantial and if justice is better served with the enforcement of the promise, then the promise may be enforced despite its lacking consideration. Generally, charitable contributions are not legally enforceable, as they lack consideration and are typically deemed as gifts. However, should promissory estoppel exist (such as the charity detrimentally relied on the contribution for some specific purpose), then the promise may be enforceable.

5.5. Capacity and Legality

Contracts exist between two or more persons with legal capacity, generally two or more mentally competent adults (though minors can enter contracts, they generally have the right to disaffirm any contract except for necessaries of food, clothing and shelter). Generally, in the business world capacity is presumed since most business is conducted by legally competent adults.

The contract's subject matter must be that of a legal one to be enforceable. For example, a contract for the sale of heroin would not be enforceable due to the illegal nature of the subject matter. However, there are some cases in which the majority of the contract might be legal and only parts of it are illegal. The courts can choose to enforce the legal parts and reject or void the illegal parts as long as doing so would not affect the essence of the bargain. In this

case the contract is considered severable and the illegal parts of the contract can be "severed." The only types of contracts that are contrary to public policy, and yet still enforceable, are: contracts in the restraint of trade or covenants not to compete (anticompetitive agreements).

5.6. Statute of Frauds

The statute of frauds basically states that certain types of contracts must be in writing and signed by one or more of the parties to be enforceable (at the very least, the party to be charged with performance must have signed the contract and the other party can then generally enforce the contract even if she did not also sign it). Contracts that are governed by the statute of frauds include: interests in land, performance cannot be completed within one year, collateral or secondary contracts, promises in consideration of marriage, sale of goods under the UCC for goods priced $500 or more. The statute of frauds is directed to the form of the contract and is a defense to the enforcement of a claimed contract.

5.7. Genuineness of Assent

Another significant defense to the enforceability of a contract is the lack of genuineness of assent; both parties must genuinely and voluntarily intend to be bound by the contract as defined by its terms. There may be a lack of genuineness of assent due to a mistake, misrepresentation, duress or undue influence. Mistakes can take the form of either bilateral or unilateral mistakes of fact and/or mistakes of value.

A bilateral, or mutual, mistake arises when both parties mistake a material fact. Should a bilateral mistake occur, either party could choose to rescind or cancel the contract. A bilateral mistake can also occur if the parties attach a different meaning to a term, material to the contract. Either party, in this case, would still be able to rescind or cancel the contract. However, should one of the parties mistake a material fact in a unilateral mistake, the party is not entitled to any relief from the contract. There are, however, two exceptions to this rule: if one party knew of or should have known of the mistake, then the contract may not be enforceable. The second exception is that of mathematical errors made inadvertently and without negligence. Should this second exception occur, such as in a mistake in addition, subtraction, division or multiplication, then the contract may be rescinded or cancelled. In the case of mistakes of value, where the mistake concerns a future market value or quality of the contract's subject matter, the contract may typically be enforced by either party. Due to the subjective nature of "value," mistakes of value hold no legal significance.

Fraudulent misrepresentation occurs when a material fact is misrepresented, one party has the intent to deceive and the innocent party has justifiably relied on the misrepresentation, and thereby resulting in injury. An innocent party fraudulently induced to enter a contract is able to rescind the contract and be restored to his or her original position or to enforce contract and seek damages for injury as a result of the fraud. An innocent party may only be able to collect damages if injury has been incurred. Note that a representation of future facts or predictions is not considered a fraudulent misrepresentation and is thereby not subject to claims of fraud. And yet, fraudulent misrepresentations can occur by conduct. For example,

> Suppose that Boileon wants to sell you a "like new" car that has "never been damaged." However, Boileon has put a cheap resin fiberglass compound on the front fenders of the car to cover up the car's prior damage. Boileon does not tell you this car was ever in an accident. Because Boileon has covered up the damage and not informed you of the prior damage and the makeshift repair, it would be considered a fraudulent misrepresentation and concealment.

Misrepresentations of the law, however, do not entitle a party to relief because parties are assumed to know state and local law. An exception to this rule is when that misrepresentation is by a person of profession known to require more knowledge of the law than an average person, such as a lawyer. Silence can usually not be construed as fraudulent misrepresentation unless the seller fails to disclose a serious material fact that the seller is under an obligation to disclose that may drastically alter the buyer's decision; for example:

> Suppose that you were planning to buy a pre-owned vehicle. However, two years prior the car had been in an accident and although the exterior fiberglass and paint had been restored to its "original look," there was major damage to the car's frame; weakening it and potentially seriously impairing the overall safety of the automobile. This concealment of important information would be considered a fraudulent misrepresentation.

5.8. Performance and Remedies

The primary reason for contracts is to benefit from the resulting performance; people make contracts to get what they want or need.

Most contracts are absolute promises in which performance must occur or else the promising party is in breach of contract. However, sometimes performance is contingent upon an event, or condition. A condition is a possible future event that triggers the performance or termination of a legal obligation. Should the condition(s) not be satisfied, the legal obligations will be discharged, be suspended, or otherwise change.

There are three types of conditions: conditions precedent, subsequent, and concurrent. Condition precedent requires the condition is fulfilled prior to performance. Conditions subsequent terminate a party's promise to perform and concurrent conditions occur when each party's absolute duty to perform is contingent on the other's simultaneous performance. Conditions can either be express or implied.

Time is a condition, express or implied. Time is generally not of the essence in most contracts even though the parties so state. The reason why timely performance is essential must be clearly stated in the contract or else mere delay in performance, though technically a breach is not actionable in the absence of showing damages resulting from the delay.

Most contracts are performed to the full or substantially full satisfaction of the parties. Thus, the majority of contracts are discharged by performance;

the contract comes to an end when both parties have fulfilled their duties. Performance can either be complete or substantial. Complete performance can only be achieved if all terms are satisfied; however, substantial performance need only be performed in good faith to enforce the contract against the other party. If one of the parties willfully fails to complete full performance, the terms will not have been substantially performed in good faith and he or she is this subject to breach.

There are several ways in which a contract may be discharged. As with many elements of contractual formation and development, the contract can be discharged by agreement. Parties can agree to discharge the contract in and of itself, or can agree to form a new contract with the express purpose of discharging the original agreement.

A contract may also be discharged by a mutual rescission, which takes place when both parties make an agreement, which satisfies legal requirements for a contract. A rescission allows for both parties to return to their original positions prior to the agreement.

For a novation to discharge a contractual obligation, both of the parties agree to substitute a third party or one of the original parties. To form a novation, courts require (1) a previous and valid obligation, (2) an agreement between all parties to the new contract, (3) discharge of the previous party, and (4) a valid new contract.

A discharge can also occur by means of a compromise, or settlement agreement, occurring when a genuine dispute over contractual obligations arises. The agreement will substitute for the contract, expressly or impliedly discharging the prior contractual obligations.

When parties agree to accept performance different than performance originally contracted for, it becomes known as discharge by accord and satisfaction. Much the same a settlement agreement, accord and satisfaction substitute as a means of discharge, except, unlike a settlement agreement, the substitution is in the form of performance, rather than agreement.

When circumstances become uncertain, discharge of contractual obligations may arise due to operation of law. Such a release of duties can manifest in a number of ways, including: alteration (a material alteration of the contract without consent); statute of limitations (a delay in filing suit bars the availability of remedy); bankruptcy (discharging the debtor's contractual obligations); and the impossibility or impracticability of performance (via death or incapacitation, illegality of the performance, destruction of the subject matter, other obstacles that may make performance highly impractical).

The general remedy for breach of contract is compensatory damages. Compensatory damages are determined by the difference between the performance as required by the terms of the contract and the performance that is actually rendered.

Compensatory damages are limited to the value of the contract itself, plus the incidental costs of covering or curing the breach of the contract by obtaining the same performance from some other party.

Consequential damages are damages beyond the contract itself caused by the breach of the contract and for which the breaching party had expressly promised to pay in the event of breach. Consequential damages can be very costly and therefore contracting parties do not willingly agree to be liable for them. Express exclusion of consequential damages is far more common.

Punitive damages cannot be obtained under contract law. However a fraud perpetrated under a contract may also be pursued as a tort claim for which punitive damages may be awarded. However, punitive damage awards are rare and that is why such an award sometimes is front-page news. Most punitive damage awards are later set aside or substantially reduced by post-trial motions or appeals, but the reductions weeks or months later rarely make the front pages; people therefore tend to have a falsely skewed sense of punitive damages.

Certain equitable remedies are also available in contract matters. Equitable relief includes: rescission, which sets aside the contract and puts the parties back in their original positions prior to making the contract; reformation, which fixes a demonstrable mistake in the contract; and injunction, which can prohibit one or more parties from taking certain action under or contrary to a provision of the contract.

6: Contract Development

By John W. Tulac

Contracts don't write themselves! Just like short stories, magazine articles, or technical reports it can be hard work to prepare a well written contract. This article provides you some guidelines to use in the preparation of written contracts.

6.1. Some Principles

When preparing any type of contract an excellent principle is to develop the relationship then build the deal. After all, the contract is supposed to document the relationship two or more parties wish to have. It only makes sense to understand that relationship before you make the deal.

The Rotary Four-Way Test is good place to start. Ask the following questions about the proposed contract.

1. Is it the Truth?
2. Is it fair for all concerned?
3. Will it build good will and better relationships?
4. Will it be beneficial for all concerned?

Many people think the goal of negotiating a contract is to gain an advantage over the other party or parties in the contract. However, this is a recipe for future performance problems, disagreements, and litigation. The concept of Principled Contracting strives for balance and fairness among all parties. Principled contracts are much more likely to be successfully executed, than one-sided contracts. Here are some Rules of Thumb for Principled Contracting.

1. Know your counterpart
2. Say what you mean and mean what you say
3. Keep it simple, stupid
4. Balanced contracts get performed. Inequitable contracts don't.
5. If the subject matter is important to you, get it in writing.
6. Do more than the contract requires.
7. Know when to walk away

6.2. Building Contracts by Asking Questions

Questioning is a good technique for developing principled contracts. Ask simple questions when developing contracts. Questioning is also useful when applying the IRAC methodology. Keep asking questions until you have all the answers.

Here are some good questions to ask when developing contracts for goods.

1. What do you want?
2. How many do you want?
3. When do you want them?
4. Where do you want them?
5. What happens if they are late?
6. How much is it/are they?
7. When is payment due?
8. In what form is payment due?
9. What happens if ____?
10. Who is responsible if___?

Here are some good questions to ask when developing contracts for services.

1. What do you want?
2. What do you already have?
3. When do you want it?
4. Where do you want it?
5. What happens if something is late?
6. How much will the work cost?
7. When is payment due?
8. In what form is payment due?
9. What happens if ____?
10. Who is responsible if___?

If you think about it, these ten questions embody the four professional requirements of a contract: Scope, Schedule, Cost, and Terms & Conditions.

Here are some good questions to ask when applying the IRAC methodology.

1. Who is/are the Plaintiff(s)? Who is/are the Defendant(s)?
2. What is/are the claim(s), causes of action?
3. What are the elements of each claim/cause of action?
4. What are the defenses to this claim/cause of action?
5. Is each element proved? What is the evidence for each element?
6. Is the defense proved? What is the evidence for each defense?

These principles and questions are useful tools in the development of contracts. However, it's often difficult to ask and answer these questions at the start of developing a contract. It often helps to have a guide or template to start from.

6.3. A Sample Contract Development Guide

In the real world developing contracts for professional services are not simple. There are many contract guidelines and sample contracts available for the design profession. One such guide, developed by Matthew Bender & Company, Inc., is presented on the following pages. While this guideline may be overly simplistic, it will give you an idea of how to start contract organization and drafting.

Volume 28: Performance of Services Chapters 84-87A
VI PERFORMANCE OF SERVICES
Chapter 86 AGREEMENTS FOR PROFESSIONAL SERVICES
Part III TRANSACTION GUIDE
C Drafting Guide

28-86 California Legal Forms--Transaction Guide § 86.140

§ 86.140 Drafting Guide for Professional Services Agreements
NOTE: In approaching the task of drafting an agreement for professional services, counsel should first ascertain the client's desires with respect to the basic subjects the agreement should cover. Sections 86.100 and 86.101 list the basic subjects common to most contract situations. For discussion of specific issues to consider in the drafting of particular agreements, see §§ 86.120-86.127, as well as the relevant forms.

[1]--Introductory Matters

1. Specify date and place of agreement.
   NOTE: The date and place is commonly set forth either at the beginning of the agreement or immediately above the signature block as "Executed on _____ [date], at _____ [place]."
2. Include designation of parties and their addresses.
   NOTE: This information is commonly set forth in an introductory paragraph at the beginning of a contract [see § 86.200[2]].
   a. Name of each party.
   b. Address of each party.
   c. Designation of parties as "client" or "corporation" and "consultant."
   d. Description of client as corporation, partnership, or sole proprietor.
3. Consider including recitals of such matters as:
   a. Client's trade or business.
   b. Client's reasons for engaging professional.
   c. Professional's special skill, background, or expertise.
4. State that client engages professional and professional accepts engagement [*see* § 86.200[2], § 2.01].
   NOTE: This type of statement is generally included in the provision specifying the term of the contract, but may be separately stated.

[2]--Term of Contract

1. If contract is to be for a specified period:
   a. State that client engages professional and that professional accepts engagement for specified period [*see* § 86.200[2], § 2.01].
   b. Specify dates on which period of engagement begins and ends, if applicable [*see* § 86.200[2], § 2.01].
   c. If contract is to be renewable for additional periods, specify method, length of renewal period, and conditions to which renewal is subject, *e.g.* [*see* § 86.200[2], § 2.02]:

(1) Automatically.

(2) For succeeding terms of stated length.

(3) Subject to specified condition (e.g., unless either party gives notice to the other within specified time before expiration of any term of intention not to renew).

d. Consider providing that contract is subject to earlier termination under other provisions [*see* § 86.200[2], Article 15].

2. If contract is to perform particular task or tasks, state that client agrees to engage professional and that professional accepts engagement, but do not specify any period [*see, e.g.,* § 86.214[2] (accountant's preparation of tax returns)].

[3]--Professional Services

1. Set forth title of professional [*see* § 86.200[2], § 1.01].

2. Describe professional's duties under contract:

a. In general terms; or

NOTE: Describing professional services in general terms gives the client the flexibility to change the job duties without necessitating a change in the contract.

b. With greater particularity.

NOTE: A detailed description of the services to be provided is especially important when the contract is on a flat fee basis, in order to reduce the possibility of confusion over what is or is not included in the flat fee.

3. Specify services not to be provided by professional, if appropriate.

NOTE: A description of what is not included in the professional's services can be as important in some cases as what is included. For example, most engagement letters for income tax preparation state that handling an audit based on the tax return is not included in the preparation fee [see § 86.214[2]]. As in the case of the description of included services in P 2, above, specificity is particularly important when the professional will charge a flat fee.

4. Specify that professional's performance may be conditioned on receipt of certain information, approvals, and reports from the client on a scheduled basis [*see* 86.214[2] (accountant's preparation of tax returns)].

5. Set forth limitations on professional's exercise of duties:

a. Specify any matters requiring consent of board of directors, if applicable [*see* § 86.200[2], § 1.03].

b. Specify that professional's exercise of duties is subject to direction of officer and or board of directors, if applicable.

6. Consider stating name or title of person to whom professional reports.

7. State place where duties are to be performed, if applicable.

8. Consider including provision that professional is authorized to use the listed equipment, supplies, and/or facilities to the extent owned by another, e.g., a university [*see* § 86.202[2], P 2].

9. Consider including provision requiring professional not to engage in any competing or conflicting business pursuits or any business, commercial, or professional nature services of the type provided to client without client's prior consent [*see* § 86.200[2], § 7.01].

10. Consider use of provision requiring professional's performance to be rendered to client's satisfaction.

11. Consider including agreement that professional has neither right nor authority to contract on behalf of client without client's written consent.

[4]--Licensure

1. If professional must be licensed or certified to perform his or her job, consider including provision requiring current licensing or certification as a condition of continued engagement.

2. Consider specifying that professional will pay licensing fees and costs of any mandatory continuing education associated in maintaining license.

[5]--Compensation

1. Specify professional's compensation:

NOTE: A periodic retainer fee may be agreed on, usually in connection with one of the following methods of compensation.

a. Per hour [*see* § 86.200[2], § 3.01].

NOTE: When charges are per hour, an additional charge is normally made for the out-of-pocket expenses, including telephone and mailing or courier costs, and special reports and services received from third parties [see § 86.300[2], § 3.02].

b. Flat fee per job [*see* § 86.214[2] (accountant's preparation of tax returns)].

NOTE: A maximum or minimum fee, or both, may included in the agreement to protect the client from excessive charges and to assure the professional that the project will be worth the effort to undertake.

c. On commission.

d. Contingency fee.

NOTE: Use of contingent fees by a licensed accountant is not ordinarily permissible [see 16 Cal. Code Reg. § 62 ; § 86.16[8]].

e. As a percentage of cost or dollar volume involved.

NOTE: Financial counselors in connection with the amount of debts to be prorated often use this method. A stated maximum or minimum fee or both may be used in connection with this method of compensation.

2. Specify payment schedule.

[6]--Reimbursement of Expenses Incurred by Professional

1. Consider including provision for reimbursement of business expenses:

NOTE: If the agreement is for professional services outside the United States, the business expense clause should state whether reimbursement is to be made in United States dollars or local foreign currency.

a. Specify what expenses are reimbursable.

(1) Consider defining ''business expenses'' as those specified in agreement or as those that are deductible business expenses under the Internal Revenue Code [*see* § 86.200[2], § 3.03]; or

(2) Specifically define what constitutes reimbursable expenses [*see* § 86.202[2], § 3.02].

b. Require professional to maintain substantiating records and submit them to client [*see* § 86.200[2], § 3.03].

2. Consider provision for payment to professional of specified amount periodically to cover business expenses (as alternative to provisions for reimbursement of business expenses).

[7]--Property Rights of Parties

1. If professional is hired to perform research and development work, include provision [*see* § 86.200[2], Article 4]:

a. Assigning inventions of professional to client; and/or

b. Excluding specified inventions from coverage of agreement for services.

2. If desired, include provision for payment of royalties to professional on inventions assigned to client:

NOTE: This type of provision may be used in conjunction with a provision that gives the client the exclusive property right in inventions conceived by the consultant without providing for royalties.

For an example of this type of provision in the employment context, see Ch. 85, Employment Agreements, § 85.431.
  a. Percentage royalty; or
  b. Varying percentage royalty.
3. If professional retains ownership of inventions, consider provision giving client perpetual royalty-free license to use those inventions.

  NOTE: For an example of this type of provision in the employment context, see Ch. 85, Employment Agreements, § 85.432.

4. If professional will have access to client's trade secrets, include provisions restricting professional's use of those secrets:

  a. State that during course of engagement, professional will have access to and become acquainted with client's trade secrets, and include general description [*see* § 86.200[2], § 5.01].

  b. Specify that all files, records, documents, and other items pertinent to the particular contract situation are and remain client's property and limit professional's ability to remove them from client's premises [*see* § 86.200[2], § 5.04].

  c. Set forth professional's promise and agreement not to misuse, misappropriate, or disclose client's trade secrets [*see* § 86.200[2], § 5.02].

  d. Include professional's agreement not to engage in unfair competition with respect to client's trade secrets [*see* § 86.200[2], § 5.03]:

  (1) Professional's acknowledgment and agreement that sale or unauthorized use or disclosure of client's trade secrets constitute unfair competition.

  (2) Professional's promise and agreement not to engage in any unfair competition with client either:

  (a) During term of agreement; or

  (b) At any time thereafter.

5. Prohibit competitive activities on part of professional against client during term of engagement [*see* § 86.200[2], § 7.01].

6. Include provision prohibiting professional from engaging in unfair competition after termination of engagement [*see* § 86.200[2], § 7.01].

[8]--Termination of Agreement

1. Consider use of provision for termination on occurrence of stated event, such as [*see* § 86.200[2], § 15.02]:
  a. Breach of duty.
  b. Neglect of duty.
  c. Continued incapacity to perform.

2. Consider use of provision for termination on default, such as [*see* § 86.200[2], § 15.03]:

  a. Corporation's failure to pay professional any compensation due within 30 days after written demand for payment.

  b. Consultant's failure to complete services specified in agreement.

  c. Consultant's material breach of any representation or agreement contained in contract.

  d. Corporation's material breach of any representation or agreement contained agreement.

3. Include provision for termination on notice from terminating party to other party [*see* § 86.200[2], § 15.01].

[9]--Effect of Breach of Contract

1. Consider use of provision stating that [*see* § 86.200[2], § 11.04]:

  a. Services to be performed are of special character that gives them peculiar value.

  b. Loss of services cannot be reasonably or adequately compensated in damages in action at law.

c. Professional expressly agrees that client is entitled to injunctive and other equitable relief to prevent or remedy breach of contract by professional.

2. Specify consequences of client's breach.

[10]--Arbitration or Other Resolution of Disputes

1. Consider use of provision under which either party may request arbitration of matters in controversy [*see* § 86.200[2], § 17.02].

2. Alternatively, consider use of provision for compulsory arbitration of matters in controversy.

3. Consider providing for other forms of alternative dispute resolution, such as mediation.

> NOTE: For detailed coverage of arbitration, mediation, and other types of alternative dispute resolution, including forms of contractual provisions for each type, see CALIFORNIA FORMS OF PLEADING AND PRACTICE, Chs. 30-38 (Matthew Bender).

[11]--General Provisions

> NOTE: For more detailed discussion of standard contractual provisions of the type noted in this subsection, see Ch. 75, Formation of Contracts and Standard Contractual Provisions.

1. Specify manner of giving notice to other party [*see* § 86.200[2], § 17.01].

2. Provide for attorneys' fees and costs [*see* § 86.200[2], § 17.03].

3. State that contract contains entire agreement of parties [*see* § 86.200[2], § 17.04]

4. Provide for modification of contract only by subsequent written agreement [*see* § 86.200[2], § 17.05].

5. Specify effect of any waiver [*see* § 86.200[2], § 17.06].

6. Provide for event that part of contract is declared invalid, either [*see* § 86.200[2], § 17.07]:

   a. Remaining provisions of contract to continue in full force and effect [*see* Civ. Code § 1670.5(a) (if contract contains unconscionable provision, court may sever or limit application of that provision, and enforce remainder of contract)].

   b. Invalid provision to be deemed amended as court or arbitrator finds reasonable.

   > NOTE: Generally, the courts lack the power to reform unenforceable contract provisions (as opposed to merely severing or limiting them). However, in the context of an employment agreement, one court of appeal held that if an agreement contains a clause in which the parties expressly agree to allow the court or an arbitrator to sever or modify any unreasonable or unenforceable provisions, the arbitration provision as so modified may be enforceable [Blake v. Ecker (2001) 93 Cal. App. 4th 728, 743-746, 113 Cal. Rptr. 2d 422].

7. State that California law governs agreement [*see* § 86.200[2], § 17.08].

8. Include date and signature blocks.

7: Business Planning and Marketing

By Abdul Rashidi and William A. Kitch

Every organization whether in manufacturing, retail, or professional service, requires a business plan. The plan defines the objectives and nature of the business, as well as strategies and an implementation plan for carrying out the strategies. The business plan must be prepared in a clear and concise manner so that it becomes the roadmap for the operation and success of the business.

A business plan typically includes and addresses, in detail, the following major components:

1. The organization plan
2. The marketing plan
3. The financial plan
4. Project Analysis (for engineering services)

While the above components are necessary and essential for an organization, the marketing plan is arguably the most critical component. The marketing plan is the most dynamic and costly component in a business plan. It is a leading element in the business plan. Successful marketing allows an organization to grow and to be profitable. Inadequate or misdirected marketing, on the other hand, results in shrinking and can even business failure. In 2009 we witnessed two of the three US auto companies (a historic business giant) crumble and undergo bankruptcy because of their inability to predict future market desires (buyers desires) and their appetite for high-efficiency, low-cost, and highly-reliable cars. In contrast, the successful marketing plan and execution of the rival foreign auto industry helped them survive the financial crises of 2008-2009 and continue their high-level sales in the US.

This article will discuss all four elements of the business plan. However, the bulk of the contents will cover the marketing plan because of the key role it plays in business planning. The financial planning section of the business plan will be expanded in Article 8.

7.1. The Organization Plan

The organization plan describes the nature of the business, long-term strategies, business operating model, and it may include the following elements:

1. Description of the Business

This element includes the type of business, a mission statement, and a business model and strategy which should include both short-term and long-term objectives.

2. Type of Products or Services

For a manufacturer, this element provides a description of the products, availability and distribution of the products, and other relevant information. For a professional service provider, it provides a description of the current and future products, general and unique special services.

3. Geographic Location

This element provides an overview of the company and describes the company headquarters and current and projected locations. It not simply a list of operating locations, but should identify the legal and geopolitical regions in which the company operates.

4. Organization Legal Structure

This element describes the company's legal structure (e.g., corporation, partnership, etc.) and why the selected legal structure is advantageous for this company.

5. Management Structure

This element describes how roles and responsibilities are divided among the key leadership in a company. It should include an organization chart, a listing of owners and/or corporate officers, the roles and responsibilities key personnel, and often salaries. It may include resumes to show the management strengths of key personnel.

6. Human Resources

This element provides current and future number of employees and positions plus requirements for level of education, professional and experience qualifications. It may also include budgetary information on salaries and benefit costs.

7. Other Elements

The organization plan also include a number of other elements which provide information on the accounting system used, the accounting company employed, monthly or periodic financial analysis, preparation of profit and loss statements, tax preparation, attorney services for legal matters, type and cost of insurance and the insurance carrier used.

7.2. The Marketing Plan

The essence of marketing is to identify current and future needs and desires in the marketplace and then align a company's products, skills, and resources to fulfill those needs and desires. The basic understanding of marketing is important not only to engineers and engineering managers but to everyone working in

a company. If the marketing process is accomplished efficiently, it will make everyone better off by creating strong opportunities and challenges, a higher standard of living, enhanced life style, etc. With the advancement in technology, improved world economy, globalization and higher standard of living throughout the developed countries, the importance of marketing has increased during the past decade and will continue into the future.

One might ask: why marketing? Here are a few reasons:
- Marketing is required and without it cannot depend on future work
- Marketing does cost money, but the money spent is an investment for future income—as the saying goes, there's no free lunch
- Successful marketing requires in-depth planning—need to know where and why we're spending money on marketing
- Marketing is everyone's job, everybody does it—even the new staff engineer

7.2.1. Elements of Marketing

Many views and opinions are offered and discussed in the literature about functions of marketing. Although there are many different approaches to marketing they all contain many of the same key principles. The simple approach to key elements of marketing offered by Hise[1] is discussed here. These key elements and their brief descriptions are as follows:
- Product advertising
- Product pricing
- Collecting product market information
- Product management
- Product physical distribution
- Product distribution channel management
- Personal selling

1. Product Advertising

Product advertising is concerned with informing the users or buyers regarding the availability of services and goods so they can purchase them to fulfill their needs and desires. In product advertising, the important duties performed are advertising budget determination, media selection for advertising, advertising message preparation, and advertisement effectiveness monitoring and testing.

2. Product Pricing

Product pricing is concerned with pricing products and services. People involved with pricing, generally management, will make many decisions, such as price objectives, basic price setting procedure, price changes procedure, discount policy, establishing product prices, etc.

3. Product Market Information

Product market information is collected by the marketing research unit of a company and is provided to the decision makers. This information deals with finding out market needs, buyer's quality desires, and satisfaction with the current products and services. The information is typically collected on the following items:

- Size of population and trends in areas under consideration
- Size of potential customers
- Potential sales volume
- Identification of profitable and unprofitable segments of the market
- Types of buyers
- Purchasing decision makers
- Product price range
- Competition
- Types of advertising
- Company image in the market

4. Product Management

The main responsibility of this element is to develop new products and services which meet and exceed the requirements of the buyer. This task or responsibility is usually accomplished through the collaboration of the production and the research and development departments of the company. Example decisions include: types and quality of product, warranties, the brand name, product support in the field, etc.

5. Product Physical Distribution

This element deals with the manufactured goods movement from the plant to the end users. The manufactured goods transportation and storage are the two major items involved in this element. In the physical distribution area, warehouse types, locations, and sizes are important. The other duties associated with this task are making decisions regarding customer service, optimum number of items in the inventory, mode of transportation, shipment size, etc.

6. Product Distribution Management

In the manufacturing goods industry, the companies generally use middlemen such as wholesalers and retailers to market their produced goods. In this case, the wholesalers and retailers are the distribution channels. Major duties of this function are making decisions on whether or not to use middlemen, selecting middlemen, performing distribution channels effectiveness analysis, etc.

However, in the professional engineering business, middlemen are not used and mass distribution is not applicable. Selling and marketing is performed by designated company individuals, such as the company office manager and project managers.

7. Personal Selling

Personal selling is similar to the aforementioned product advertising element but the difference is that personal selling makes use of personal resources to inform customers and clients about the existence of products which can satisfy their needs. In other words, the marketing personnel call on customers and clients. This method is widely used in marketing professional engineering services.

7.2.2. Marketing Approach

An effective marketing plan is a process which starts with an in-depth viability analysis of the market, assessing the needs and desires of the buyers, setting objectives for product development, developing strategies for sale of product, and finishes with a feedback control system to ensure compatibility between the product quality and buyers' needs. It is also an interactive process, so that the output of each stage is checked to see what impact it may have on the earlier stages and to modify as necessary.

The value of this process is that it moves from the general to the specific; from the overall objectives of the organization down to the individual action plan, with a visible audit trail.

1. Perform Market Analysis

Market analysis identifies the needs of the market. Once the needs are known, the company can decide on the best plan to serve the identified market, etc.

Marketing analysis focuses on:
- Looking in—who are we, what do we do?
- Looking back—what have we done well in the past?
- Looking out—who are we working for?
- Looking ahead—what changes are coming?

An analysis of the past leads to an understanding of the future. Therefore, trend analysis is often performed to prepare predictive future business models. These models serve as the 1, 5, and 10-year business and marketing plans.

A typical example of a market analysis for the Hotshot Structural Engineering firm is presented in Figure 7.1. This analysis of the last fiscal year's income and revenue indicated that this firm added 5% new client for 15:1 revenue/cost ratio – is this okay? While the revenue/cost ratio for the 5% new client is relatively high for this firm, however the added client justifies the cost.

2. Perform Environmental Analysis

Environmental analysis is concerned with evaluation and assessment of the competition, health of economy, availability of technology, legal aspects, cultural issues and others.

3. Establish Product Objectives

The next step is to establish objectives. A firm usually set two types of objectives:

(1) Strategic objectives - The strategic objectives are long term, taking more than one year to accomplish, and are the core objectives of the firm. They are established early on as part of the organizational planning activities, and include such items as volume of sales, % annual growth, % share of the market, % profit and geographical locations. These objectives are set by the company owners and upper management.

(2) Tactical objectives - The tactical objectives are generally short term, often one year long, and are assigned to the firm's marketing and other departments.

4. Achieve Strategic Objectives

In this step, through several combinations of products and markets, the marketing and management personnel make decisions relative to achievement of strategic objectives. The four different product/market combinations used are presented in Figure 7.2.

5. Achieve Tactical Objectives

This step is concerned with accomplishing tactical objectives. It starts with first translating the high-level strategic objectives into lower-level tactical objectives and then developing specific action plans for implementation. In marketing professional services, the strategic objectives are first translated into regional and office-level strategies and then refined to specific-project-level action plans. Examples are: switching the advertising media and updating brochures to increase the number of potential users, offering seminars to increase clients/customers product knowledge, developing a strategic marketing plan for winning a project, etc.

A typical marketing/business plan may include the following product line categories and segment classifications. This serves as a useful tool for both marketing and budget and project management analysis purposes.
- Marketing Product Area—major product areas which a firm is performing business in and delivering their products
- Marketing Segments—types of products within a product area
- Marketing Sub-segments—types of products within a segment area

- Marketing Unit—projects within a sub-segment, segment and/or a product area

An example of the business plan for MWW, Inc. Environmental Engineering is presented in Figure 7.3.

6. Setup Marketing Organization

The next step is to set up the best marketing organization to achieve these objectives. Marketing organization is concerned with the basic organizational approach, determining the span of control, degree of specialization, centralization/decentralization of product development, size and training of marketing personnel, and procedure for motivating marketing persons.

7. Establish a Feedback Control System

The objective of the feedback control system is to monitor the performance of the marketing activities and provide feedback on their effectiveness. To conduct this activity, the company should establish the monitoring aspect of the marketing operation, the measuring criterion, the performance standards.

7.3. The Financial Plan

Financial plan includes all financial documents and records that indicate past, current, and future finances. This section of the business plan quantifies everything that was stated in the organizational and marketing plans. For this reason, this part of the business plan is done following completion of the other two parts.

Major documents that will be included in the financial plan are briefly described below. The information should be developed and organized in the order presented below because they build on each other, utilizing information from the ones previously developed.

1. Financial Needs

A financial need statement is required if the business is seeking financing and loans. It should cover the purpose and amount of loan needed and how the funds will be used.

2. Budget and Cash Flow Analysis

This document interprets the business plan requirements and expresses them in terms of dollars and provides cash inflow and outflow projections over a period of time. It is of prime interest to the lender and shows how the business intends to financially operate and to repay the loan.

3. Income Projections

A business prepares a Income Profit and Loss Statement (P&L Statement) for each operating year. This statement is typically prepared at the beginning of a fiscal year covering the previous 12 months of operation. Utilizing the records of the past year and the budgetary requirements of the marketing plan for the next 3 to 5 years, income projections are then prepared.

4. Profit and Loss Statement

The profit and loss statement covers a company's financial activity over a period of time (monthly, quarterly, annually). It provides a moving picture showing what has happened in the business and it is an excellent tool for assessing business health.

5. Balance Sheet

A balance sheet is a summary of Assets, Liabilities, and Net Worth of the company as of a certain date. It gives a picture of the firm's financial conditions at a particular time and shows whether a firm's financial position is strong or weak. Balance sheets may also be projected for the current or coming years utilizing the budget and income projections.

6. Financial Statement Analysis

Utilizing income statements and balance sheets, it is easy to develop a comparative analysis of several years financial statement to study trends for overall company business growth/decline, business market sector growth/decline, comparison of this business with competition, P&L trends, etc. Results are expressed as ratios or percentages that can be used to compare this business with industry standards. These analyses are also important documents in determining whether or not a loan is justified. Typical analysis include: liquidity analysis, return on investment, debt ratios, profitability analysis, etc.

7. Financial History

This is a summary of financial information of a company from its start to the present.

7.4. Project Analysis

Project analysis is a process of collecting relevant existing information on company projects and conducting trend analysis in an effort to develop a predictive business model. The model will assist the company decision makers to make realistic and informed decisions relative to the strategic objectives and goals.

A typical project analysis may include the following:
- Develop a project analysis form (see attached PAF)
- Gather records (technical, performance and financial) on all projects done in the last 5 years
- Complete the PAF for each project
- Develop a project analysis spreadsheet from the PAFs
- Develop criteria for analysis
- Perform analysis and present the results both in tabular and graphical formats
- Research, review, and document future (1, 5 and 10-year horizons) market potentials, market shares, volatility, and variability in the company's established product lines
- Develop trends and predictive models for the business plan

7.5.   References

1. Hise, R.T., P.L. Gillette and J.K. Ryans; *Basic Marketing: Concepts*. Cambridge, MA, Winthrop Publishers, Inc. 1979

Annual Sales Revenue/Cost Presentation
Hotshot Structural Engineering, Inc.

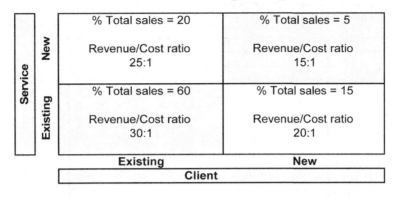

Figure 7.1: Typical marketing analysis

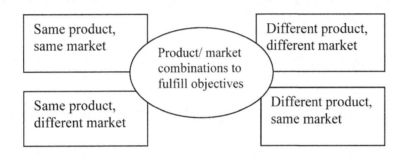

Figure 7.2: Product market combinations to fulfill strategic objectives

MWW, Inc. Environmental Engineering
General Marketing and Planning Categories

Product line ⇨	Planning	Design	Construction Management	Design Build Construction
Segment ⇨	- Environmental Studies - Master Plans	- Water - Wastewater	- Water - Wastewater	- Water - Wastewater
Subsegment ⇨	- Water - Wastewater	- Plants - Pump stations	- Plants - Pump stations	- Plants - Pump stations
Marketing Unit ⇨	- Project 1 - Project 2 - Project 3	- Project 1 - Project 2 - Project 3	- Project 1 - Project 2 - Project 3	- Project 1 - Project 2 - Project 3

Figure 7.3: Business plan components

Project Analysis Form (PFA)
General Marketing and Planning Categories
MWW, Inc. Environmental Engineering

Item	Description
Year	
Project Name (MU)	
New Project	
Existing Project	
Services Provided	
Owner	
Client	
Project Location	
Project size, $	
Fee (MWW)	
Profit, %	
Problems/Issues	
Product line	
Segment	
Subsegment	

Figure 7.4: Typical Project Analysis Form showing key data that must be gathered for each project

8: Budget Planning

By Abdul Rashidi and William A. Kitch

As discussed in Article 7, the financial component of the Business Plan contains the following elements:

1. Financial Needs
2. Budget and Cash Flow Analysis
3. Income Projections
4. Profit and Loss Statement
5. Balance Sheet
6. Financial Statement Analysis
7. Financial History

Budget planning, the task of categorizing income and expenses and tracking funds in each category, is at the core of all financial planning. Without a budget it is impossible to complete the rest of the financial plan. Therefore this Article will present the budget planning process as the core of the financial plan. It will also introduce a simple profit and loss statement to demonstrate an important application of the budgeting process.

8.1. Budget Planning

Budgeting encourages a business entity to develop the most efficient strategies for being profitable and expanding business. Budgeting also help a firm to use its financial and human resources in an effective way.

Efficient budgeting process also incorporates comprehensive review and analysis of past trends and data pertinent to the business. This analysis assists a company on many facts including type of market and geographic locations, business organization structure needed, the amount of money to be invested, the marketing strategies required, and the type and number of resources to hire.

Budgeting can also provide additional benefits, such as an increased awareness of costs and revenue, monitoring of company goals, and a model for performance evaluation and changes.

A simple budget plan includes the following major components:

1. **Revenue**—the amount of money a company earns. For a consulting engineering business, the primary source of revenue is from professional services or from the sale of technical professionals' man-hours. The man-hour rates are adjusted by a factor, called Labor Multiplier Factor, expressed as % of salary, to account for salary cost, overhead, and profit. Overhead and profit are needed to run a company's business. A few important terms to note are:
   - **The utilization rate**—This is the % of an employee's time charged to a project.
   - **The hourly invoice or billing rate**—This rate includes three items: the salary cost, the overhead cost, and the profit margin. Overhead includes indirect salary cost, and other expenses for running a business, such as the payroll taxes, insurance, and general and administrative expenses. More on this topic later.
   - **Labor Multiplier Factor (LMF)**—This factor incorporates all costs (salary and non—salary) and profit into a single multiplier, and it is

expressed as a multiplier of the salary. The intent is to have adequate revenue to meet company's financial goals and objectives.

- **Revenue sources**—include: the professional services (labor man-hour), reimbursement expenses, purchased (subcontract) services, and equipment charges.

2. **Cost**—the amount of money a company spends to maintain a successful operation. Basically, there are two cost categories: Direct and Indirect costs. Direct cost includes primarily salary expenses for labor directly charged to a project. Indirect cost, also called overhead, includes: indirect labor (expenses for labor not charged to a project) and other non-salary costs.

- **Labor cost**
  - Direct labor—time spent on projects and it is chargeable
  - Indirect labor—time not chargeable to projects (office admin, marketing, attending conferences, education and training, etc)
- **Material Cost**
  - Direct Material—chargeable to projects (also called other direct costs or ODC)
  - Indirect Material—non-chargeable to projects
- **Overhead Cost**—Overhead includes: indirect labor (expenses for labor not charged to a project); payroll burden; and general and administration expenses. Overhead costs are essential to running a business operation. An example of typical overhead expenses, for engineering services, is presented in Table 1.
  - Payroll burden—amount of money provided for payroll taxes, SDI, vacation and sick leave, health insurance—Payroll burden is applied to direct and indirect labor.
  - General & Admin (G&A) costs—amount of money provided for running a business, including: rent, supplies, utilities, professional insurance, bank loan interest, equipment lease, autos, income tax, professional services, etc. Payroll burden and G&A are developed from past records, updated annually and it pays for all expenses NOT chargeable to projects.

- Total Overhead = Indirect labor + Payroll burden + G&A

3. **Profit**—Business is about making money in a professional and ethical way. Money allows business to improve and expand and to sustain competition. Profit pays for training and education, contingency reserve, self-insurance, improving and expanding business operation, retirement, employee bonuses, stockholders, etc.

Profit = Revenue – Cost

- Revenue—Income from services provided or total sale
- Cost—Cost of all labor, material, and overhead

8.2. Budget Worksheet

Budget worksheet is a simple tool to display and capture the major components of a company's business, including: revenue, cost, and profit. It is a simple model to estimate, project, and predict outcomes. It is also used to monitor and evaluate business performance, and to control budgeting process—in other words, it is a model for change.

An example of a step-by-step preparation of a budget worksheet is presented below. In this example two case scenarios are reviewed, including:

- Case 1—Labor multiplier factor is not given. Budget calculation therefore revolves around assuming employees utilization rates and invoice rates and then checking the resulting LMF and profit margin.
- Case 2—Labor multiplier factor as well as overhead factors are available from the company's last year's actual records. Budget calculation is therefore simple and easy and the goal is to adjust the staff utilization rates in doer to arrive at a reasonable profit margin, perhaps exceeding the company goal.

Note that a budget worksheet can be used for business planning and monitoring of an office as well as an individual project.

Budget Worksheet Calculation
Case 1- Labor Multiplier Factor is Not Given

Name	Title	Annual Salary ($)	Total Hours	Hrly Salary	Planned % Utilization	Direct Hours	Indirect Hours	Labor Costs		Revenue		
								Direct	Indirect	Invoice Hourly Rate	Labor Multiplier (LMF)	Revenue
Maria	Principal	$125,000	2,080	60	30%	624	1,456	$37,500	$87,500	180	3.00	112,320
Harry	Project Manager	$115,000	2,080	55	85%	1,768	312	$97,750	$17,250	170	3.07	300,560
Yousef	Project Engineer	$95,000	2,080	46	85%	1,768	312	$80,750	$14,250	140	3.07	247,520
Rebecca	Staff Engineer	$67,000	2,080	32	85%	1,768	312	$56,950	$10,050	97	3.01	171,496
Kit	Staff Engineer	$67,000	2,080	32	85%	1,768	312	$56,950	$10,050	97	3.01	171,496
Fran	CAD Operator	$42,000	2,080	20	85%	1,768	312	$35,700	$6,300	61	3.02	107,848
John	Office Mnager /Mark	$100,000	2,080	48	50%	1,040	1,040	$50,000	$50,000	145	3.02	150,800
Gaspar	Admin Asst	$30,000	2,080	14	50%	1,040	1,040	$15,000	$15,000	40	2.77	41,600
Beth	Accountant	$45,000	2,080	22	50%	1,040	1,040	$22,500	$22,500	60	2.77	62,400
		$686,000	18,720			12,584	6,136	$453,100	$232,900		Total Revenue	$ 1,366,040

Total Revenue $ 1,366,040

Ave. LMF(Rev/ Dir Labor) 3.01

Part A - Revenue Calculations:

Revenue for an engineering consulting firm comes primarily from sale of man-hours, or the number of hours charged to projects for a given period of time. To determine the annual revenue, the following factors are important:

- **The utilization rate** for all employees, or the % of their time charged to a project.
- **The hourly invoice or billing rate.** This rate includes three items: the salary cost, the overhead cost, and the profit margin. Overhead include indirect salary cost, and other expenses for running a business, such as the payroll taxes, insurance, and general and administrative expenses. More on this topic latter.

1. Prepare a spreadsheet and list all employees and their annual salaries
2. There are 2080 working hours in one year. Calculate Hourly Salary by dividing the annual salary to the total hours for each employee.
3. Assign % Utilization (this is the % of their time charged to a project).
4. Using the % utilization, calculate Direct Hours by multiplying utilization rate to the total hours.
5. Calculate Indirect Hours by subtracting direct hours form total hours.
6. Calculate Direct Cost by multiplying hourly rate to direct hours.
7. Calculate Indirect Cost by multiplying hourly rate to indirect hours.
8. Assign Hourly Rates for invoicing projects. These rates should cover the salary cost as well as the overhead cost and profit (typical profit is 10% of total cost).
9. Calculate Labor Multiplier Factor (LMF) by dividing the invoice hourly rate to the hourly salary. This number should be close to last year's average LMF (typical LMF for an engineering consulting firm ranges from 2.6 to 3.2).
10. Calculate Revenue by multiplying invoice hourly rate to direct hours.
11. Calculate Total Revenue by adding individual revenue items.
12. Calculate Average LMF by dividing total revenue to direct labor cost.

Budget Worksheet Calculation
Case 1 - Labor Multiplier Factor is Not Given

Name	Title	Annual Salary ($)	Total Hours	Hrly Salary	Planned % Utilization	Direct Hours	Indirect Hours	Labor Costs Direct	Labor Costs Indirect	Invoice Hourly Rate	Revenue Labor Multiplier (LMF)	Revenue
Maria	Principal	$125,000	2,080	60	30%	624	1,456	$37,500	$87,500	180	3.00	112,320
Harry	Project Manager	$115,000	2,080	55	85%	1,768	312	$97,750	$17,250	170	3.07	300,560
Yousef	Project Engineer	$95,000	2,080	46	85%	1,768	312	$80,750	$14,250	140	3.07	247,520
Rebecca	Staff Engineer	$67,000	2,080	32	85%	1,768	312	$56,950	$10,050	97	3.01	171,496
Kit	Staff Engineer	$67,000	2,080	32	85%	1,768	312	$56,950	$10,050	97	3.01	171,496
Fran	CAD Operator	$42,000	2,080	20	85%	1,768	312	$35,700	$6,300	61	3.02	107,848
John	Office Mnager /Mark	$100,000	2,080	48	50%	1,040	1,040	$50,000	$50,000	145	3.02	150,800
Gaspar	Admin Asst	$30,000	2,080	14	50%	1,040	1,040	$15,000	$15,000	40	2.77	41,600
Beth	Accountant	$45,000	2,080	22	50%	1,040	1,040	$22,500	$22,500	60	2.77	62,400
		$686,000	18,720			12,584	6,136	$453,100	$232,900		Total Revenue	$ 1,366,040

Total Revenue	1,366,040
Direct Costs	
Labor	$453,100
Indirect Costs (Overhead)	
Payroll Burden	$198,940
Indirect Labor	$112,500
G&A	$468,000
Total OH	**$779,440**
Total Costs (Direct + Indirect)	**$1,232,540**
Profit (Revenue - Total Costs)	$ 133,500
Profit %	**10.8%**
Average Labor Multiplier (Revenue/ Dir Labor)	**3.01**

Part B- Cost Calculations:

Basically, there are two cost categories: Direct and Indirect costs. Direct cost includes primarily salary expenses for labor directly charged to a project. Indirect cost, also called Overhead, includes: indirect labor (expenses for labor not charged to a project); payroll burden; and general and administration expenses. Overhead costs are essential to run t a business operation. An example of typical overhead expenses, for engineering services, is presented in Table 1.

1. For purposes of budget calculation, costs are further divided into Direct and Indirect components. Direct cost equals to Direct Labor cost, as calculated in the budget worksheet. Indirect costs represent the costs necessary to run the company operation. They cannot be directly charged to a project but can be included as an overhead.

2. Calculate Overhead Costs - An example of the typical overhead costs for our case project is presented in Table 1. As shown, the components costs are either calculated or based on last year's actual data. Typical total overhead cost ranges 150 -180% of direct labor cost.

3. Calculate Overhead Multiplier Factor for each overhead component by dividing that overhead cost to total direct labor cost.

4. Calculate Breakeven Overhead Multiplier Factor by dividing the total overhead cost by the total direct labor cost.

5. Calculate Labor Multiplier Factor (LMF) by adding profit, expressed as multiplier factor, to the Breakeven Multiplier. Typical LMF ranges from 2.8 to 3.2.

Part C - Profit Calculations:

1. Profit Margin- Minimum Profit goal is determined by company management as a % of the total cost. It ranges from 8-15%.

2. Profit amount is calculated as: Revenue - Total cost.

3. Profit is also expressed as % of total cost. This is done by dividing the profit amount to the total cost, and then converting it to %.

Budget Worksheet Calculation
Case 2 - Labor Multiplier Factor is Given

Name	Title	Annual Salary ($)	Total Hours	Hrly Salary	Planned % Utilization	Direct Hours	Indirect Hours	Labor Costs		Labor Multiplier (LMF)	Revenue	
								Direct	Indirect		Invoice Hourly Rate	Revenue
Maria	Principal	$125,000	2,080	60	30%	624	1,456	$37,500	$87,500	3	180.29	112,500
Harry	Project Manager	$115,000	2,080	55	85%	1,768	312	$97,750	$17,250	3	165.87	293,250
Yousef	Project Engineer	$95,000	2,080	46	85%	1,768	312	$80,750	$14,250	3	137.02	242,250
Rebecca	Staff Engineer	$67,000	2,080	32	85%	1,768	312	$56,950	$10,050	3	96.63	170,850
Kit	Staff Engineer	$67,000	2,080	32	85%	1,768	312	$56,950	$10,050	3	96.63	170,850
Fran	CAD Operator	$42,000	2,080	20	85%	1,768	312	$35,700	$6,300	3	60.58	107,100
John	Office Mnager /Mark	$100,000	2,080	48	50%	1,040	1,040	$50,000	$50,000	3	144.23	150,000
Gaspar	Admin Asst	$30,000	2,080	14	50%	1,040	1,040	$15,000	$15,000	3	43.27	45,000
Beth	Accountant	$45,000	2,080	22	50%	1,040	1,040	$22,500	$22,500	3	64.90	67,500
		$686,000	18,720			12,584	6,136	$453,100	$232,900		Total Revenue	1,359,300

In general, preparation of a budget worksheet is much easier because the cost-side data is readily available from the previous year's operation and are applied to the current- budget year.

Part A -Revenue Calculations:

1. Prepare a spreadsheet- Steps 1 to 7 is same as presented before.
2. Determine LMF from previous year's actual overhead data and planned profit margin for the budget year. For this example, the LMF is 3.0 which include a 10% profit.
3. Incorporate the LMF into this spreadsheet.
4. Calculate Invoice hourly rates by multiplying hourly salary to the LMF.
5. Calculate revenue by multiplying invoice rate to the direct labor.
6. Generate total revenue, by adding individual revenue items.
7. Determine rates for overhead components as % of direct labor, from previous year's actual overhead data and incorporate them into this worksheet. For this example, the values are: payroll = 43.9%, indirect labor = 24.8% and G&A=103.3%.
8. Calculate total indirect cost, applying the above overhead rates.
9. Calculate profit and compare it to the planned 10% profit goal. Adjust the LMF, if profit is desired.

Part B- Cost Calculations:

1. Cost calculations are much simpler as the overhead multiplier factors are determined from previous year's actual data.
2. Incorporate the overhead multiplier factors, as well as profit into the spreadsheet, and calculate total cost.

Part C- Profit Calculations:

1. Profit calculation is the same as presented before.

	%DL	Cost
Direct Labor	100.0	$453,100
Payroll OH	43.9	$198,911
Ind labor OH	24.8	$112,369
General & Admin	103.3	$468,052
Total Cost		**1,232,432**
Profit (Revenue -Total Costs)		**126,868**
Profit %		**10.3%**
Average Labor Multiplier (Revenue/ Dir Labor)		3.00

Table 8.1: Example – Establishing Overhead Costs and Rates

Payroll Burden		
	Cost	% of Total Salary
Holiday	18,795	3%
Paid Time Off (PTO)	37,589	5%
Group Health Insurance	109,760	16%
Group Life Insurance	13,720	2%
Group LTD Insurance	6,860	1%
Worker's Comp Insurance	34,300	5%
Education	34,300	5%
Professional dues and licenses	6,860	1%
Profit sharing/Pension	102,900	15%
Payroll Taxes (SS and Medicare)	54,880	8%
Other benefits	20,580	3%
Total	**$ 198,940**	

Indirect Labor		
Accounting	$22,500	3%
Marketing	$25,000	4%
Admin	$15,000	2%
others	$50,000	7%
Total	**$ 112,500**	

General and Admin (G&A)		
Rent	60,000	9%
Auto	24,000	3%
Equipment leases	48,000	7%
Maintenance /repairs	48,000	7%
Supplies	24,000	3%
Prof. Insurance	60,000	9%
Prof. Services	48,000	7%
Bank Loan interest	84,000	12%
Other Taxes	24,000	3%
Misc.	48,000	7%
Total	**$ 468,000**	

Total Overhead		
	$	OH Multiplier as % of Dir. Labor
Payroll Burden	198,940	43.9
Indirect Labor	112,500	24.8
G&A	468,000	103.3
Total Cost	**$ 779,440**	**172.0**

Dir Labor Multiplier Factor		
Direct Labor (from budget WS)	$ 453,100	100.0
Breakeven Multiplier = Direct labor + Overhead Multiplier		272.0
Profit as % total costs	10%	27.2
Labor Multiplier Factor as %		299.2
Labor Multiplier Factor		2.992

PART TWO

Readings

Attributes of a Profession

By Ernest Greenwood

The professions occupy a position of great importance on the American scene.[1] In a society such as ours, characterized by minute division of labor based upon technical specialization, many important features of social organizations are dependent upon professional functions. Professional activity is coming to play a predominant role in the life patterns of increasing numbers of individuals of both sexes, occupying much of their waking moments, providing life goals, determining behavior, and shaping personality. It is no wonder, therefore, that the phenomenon of professionalism has become an object of observation by sociologists.[2] The sociological approach to professionalism is one that views a profession as an organized group which is constantly interacting with the society that forms its matrix, which performs its social functions through a network of formal and informal relationships, and which creates its own subculture requiring adjustments to it as a prerequisite for career success.[3]

Within the professional category of its occupational classification the United States Census Bureau includes, among others, the following: accountant, architect, artist, attorney, clergyman, college professor, dentist, engineer, journalist, judge, librarian, natural scientist, optometrist, pharmacist, physician, social scientist, social worker, surgeon, and teacher.[4] What common attributes do these professional occupations possess which distinguish them from the nonprofessional ones? After a careful canvass of the sociological literature on occupations, this writer has been able to distill five elements, upon which there appears to be consensus among the students of the subject, as constituting the distinguishing attributes of a profession.[5] Succinctly put, all professions seem to possess: (1) systematic theory, (2) authority, (3) community sanction, (4) ethical codes, and (5) a culture. The purpose of this article is to describe fully these attributes.

Before launching into our description, a preliminary word of caution is due. With respect to each of the above attributes, the true difference between a

---

1   Talcott Persons, "The Professions and Social Structure," *Social Forces*, .17 (May 1939), 457-467.

2   Theodore Caplow, *The Sociology of Work,* (Minneapolis: University of Minnesota Press, 1954).

3   Oswald Hall, "The Stages of a Medical Career," *The American Journal of Sociology, 53* (March 1948), 327-336; "Types of Medical Careers," *The American Journal of Sociology, 55* (November 1949), 243253; "Sociological Research in the Field of Medicine: Progress and Prospects," *American Sociological Review, 16* (October 1951), 639-644.

4   U.S. Bureau of the Census, *1950 Census of Population: Classified Index of Occupations and Industries* (Washington, D.C.; Government Printing Office, 1950).

5   The writer acknowledges his debt to his former students at the School of Social Welfare, University of California, Berkeley, who, as members of his research seminars, assisted him in identifying and abstracting the sociological literature on occupations. Their conscientious assistance made possible the formulation presented in this paper.

professional and a nonprofessional occupation is not a qualitative but a quantitative one. Strictly speaking, these attributes are not the exclusive monopoly of the professions; nonprofessional occupations also possess them, but to a lesser degree. As is true of most social phenomena, the phenomenon of professionalism cannot be structured in terms of clearcut classes. Rather, we must think of the occupations in a society as distributing themselves along a continuum.[6] At one end of this continuum are bunched the well-recognized and undisputed professions, (e.g., physician, attorney, professor, scientist); at the opposite end are bunched the least skilled and least attractive occupations, (e.g., watchman, truckloader, farm laborer, scrubwoman, bus boy). The remaining occupations, less skilled and less prestigious than the former, but no more so than the latter, are distributed between these two poles. The occupations bunched at the professional pole of the continuum possess to a maximum degree the attributes about to be described. As we move away from this pole, the occupations possess these attributes to a decreasing degree. Thus, in the less developed professions, social work among them, these attributes appear in moderate degree. When we reach the mid-region of the continuum, among the clerical, sales, and crafts occupations, they occur in still lesser degree; while at the unskilled end of the continuum the occupations possess these attributes so minimally that they are virtually nonexistent. If the reader keeps this concept of the continuum in mind, the presentation will less likely appear as a distortion of reality.

SYSTEMATIC BODY OF THEORY[7]

It is often contended that the chief difference between a professional and a nonprofessional occupation lies in the element of superior skill. The performance of a professional service presumably involves a series of unusually complicated operations, mastery of which requires lengthy training. The models referred to in this connection are the performances of a surgeon, a concert pianist, or a research physicist. However, some nonprofessional occupations actually involve a higher order of skill than many professional ones. For example, tool-and-die making, diamond-cutting, monument-engraving, or cabinet-making involve more intricate operations than school teaching, nursing, or social work. Therefore, to focus on the element of skill per se in describing the professions is to miss the kernel of their uniqueness.

The crucial distinction is this: the skills that characterize a profession flow from and are supported by a fund of knowledge that has been organized into an internally consistent system, called a *body of theory*. A profession's underlying body of theory is a system of abstract propositions that describe in general terms the classes of phenomena comprising the profession's focus of interest. Theory serves as a base in terms of which the professional rationalizes his operations in concrete situations. Acquisition of the professional skill requires a prior or simultaneous mastery of the theory underlying that skill. Preparation for a profession, therefore, involves considerable preoccupation with systematic theory, a feature virtually absent in the training of the nonprofessional. And so treatises are written on legal theory, musical theory, social work theory, the theory of the drama, and so on; but no books appear on the theory of punch-pressing or pipefitting or bricklaying.

Because understanding of theory is so important to all professional skill, preparation for a profession must be an intellectual as well as a practical experience. On-the-job training through apprenticeship, which suffices for a nonprofessional occupation, becomes inadequate for a profession. Orientation in theory can be achieved best through formal education· in an academic setting. Hence the appearance of the professional school, more often than not university affiliated, wherein the milieu is a contrast to that of the trade school. Theoretical knowledge is more difficult to master than operational procedures; it is easier to

---

6   The occupational classification employed by the U. S. Census Bureau is precisely such a continuum. The categories of this classification are: (a) professionals and semiprofessional technical workers; (b) proprietors and managers,· both farm and non-farm, and officials; (c) clerical, sales, and kindred workers; (d) craftsmen, skilled workers, and foremen; (e) operatives and semiskilled workers; and (f) laborers, unskilled, service, and domestic workers. (U.S. Bureau of the Census, *op. cit.*).

7   The sequence in which the five attributes are discussed in this paper does not reflect upon their relative importance. The order selected has been dictated by logical considerations.

learn to repair an automobile than to learn the principles of the internal combustion engine. There are, of course, a number of free-lance professional pursuits (e.g., acting, painting, writing, composing, and the like) wherein academic preparation is not mandatory. Nevertheless, even in these fields various "schools" and "institutes" are appearing, although they may not be run along traditional academic lines. We can generalize that as an occupation moves toward professional status, apprenticeship training yields to formalized education, because the function of theory as a groundwork for practice acquires increasing importance.

The importance of theory precipitates a form of activity normally not encountered in a nonprofessional occupation, viz., theory construction via systematic research. To generate valid theory that will provide a solid base for professional techniques requires the application of the scientific method to the service-related problems of the profession. Continued employment of the scientific method is nurtured by and in turn reinforces the element of *rationality*.[8] As an orientation, rationality is the antithesis of traditionalism. The spirit of rationality in a profession encourages a critical, as opposed to a reverential, attitude toward the theoretical system. It implies a perpetual readiness to discard any portion of that system, no matter how time-honored it may be, with a formulation demonstrated to be more valid. The spirit of rationality generates group self-criticism and theoretical controversy. Professional members convene regularly in their associations to learn and to evaluate innovations in theory. This produces an intellectually stimulating milieu that is in marked contrast to the milieu of a nonprofessional occupation.

In the evolution of every profession there emerges the researcher-theoretician whose role is that of scientific investigation and theoretical systematization. In technological professions[9] a division of labor thereby evolves, that between the theory-oriented and the

practice-oriented person. Witness the physician who prefers to attach himself to a medical research center rather than to enter private practice. This division may also yield to cleavages with repercussions upon intraprofessional relationships. However, if properly integrated, the division of labor produces an accelerated expansion of the body of theory and a sprouting of theoretical branches around which specialties nucleate. The net effect of such developments is to lengthen the preparation deemed desirable for entry into the profession. This accounts for the rise of graduate professional training on top of a basic college education.

PROFESSIONAL AUTHORITY

Extensive education in the systematic theory of his discipline imparts to the professional a type of knowledge that highlights the layman's comparative ignorance. This fact is the basis for the professional's authority, which has some interesting features.

A nonprofessional occupation has customers; a professional occupation has clients. What is the difference? A customer determines what services and/or commodities he wants, and he shops around until he finds them. His freedom of decision rests upon the premise that he has the capacity to appraise his own needs and to judge the potential of the service or of the commodity to satisfy them. The infallibility of his decisions is epitomized in the slogan: "The customer is always right!" In a professional relationship, however, the professional dictates what is good or evil for the client, who has no choice but to accede to professional judgment. Here the premise is that, because he lacks the requisite theoretical background, the client cannot diagnose his own needs or discriminate among the range of possibilities for meeting them. Nor is the client considered able to evaluate the caliber of the professional service he receives. In a nonprofessional occupation the customer can criticize the quality of the commodity he has purchased, and even demand a refund. The client lacks this same prerogative, having surrendered it to professional authority. This element of authority is one, although not the sole, reason why a profession frowns on advertising. If a profession were to advertise, it would, in effect, impute to the potential client the discriminating capacity to

---

8   Parsons, *op. cit.*

9   A technology is a profession whose aim is to achieve controlled changes in natural relationships. Convention makes a distinction between technologists who shape nonhuman materials and those who deal with human beings. The former, are called engineers; the latter practitioners.

select from competing forms of service. The client's subordination to professional authority invests the professional with a monopoly of judgment. When an occupation strives toward professionalization, one of its aspirations is to acquire this monopoly.

The client derives a sense of security from the professional's assumption of authority. The authoritative air of the professional is a principal source of the client's faith that the relationship he is about to enter contains the potentials for meeting his needs. The professional's authority, however, is not limitless; its function is confined to those specific spheres within which the professional has been educated. This quality in professional authority Parsons calls *functional specificity*.[10] Functional specificity carries the following implications for the client-professional relationship.

The professional cannot prescribe guides for facets of the client's life where his theoretical competence does not apply. To venture such prescriptions is to invade a province wherein he himself is a layman, and, hence, to violate the authority of another professional group. The professional must not use his position of authority to exploit the client for purposes of personal gratification. In any association of superordination-subordination, of which the professional-client relationship is a perfect specimen, the subordinate member—here, the client—can be maneuvered into a dependent role. The psychological advantage which thereby accrues to the professional could constitute a temptation for him. The professional must inhibit his impulses to use the professional relationship for the satisfaction of the sexual need, the need to manipulate others, or the need· to live vicariously. In the case of the therapeutic professions it is ideally preferred that client-professional intercourse not overflow the professional setting. Extraprofessional intercourse could be used by both client and professional in a manner such as to impair professional authority, with a consequent diminution of the professional's effectiveness.

Thus far we have discussed that phase of professional authority which expresses itself in the client-professional relationship. Professional authority, however, has professional-community ramifications. To these we now turn.

SANCTION OF THE COMMUNITY

Every profession strives to persuade the community to sanction its authority within certain spheres by conferring upon the profession a series of powers and privileges. Community approval of these powers and privileges may be either informal or formal; formal approval is that reinforced by the community's police power.

Among its powers is the profession's control over its training centers. This is achieved through an accrediting process exercised by one of the associations within the profession. By granting or withholding accreditation, a profession can, ideally, regulate its schools as to their number, location, curriculum content, and caliber of instruction. Comparable control is not to be found in a nonprofessional occupation.[11] The profession also acquires control over admission into the profession. This is achieved via two routes. First, the profession convinces the community that no one should be allowed to wear a professional title who has not been conferred it by an accredited professional school. Anyone can call himself a carpenter, locksmith, or metal-plater if he feels so qualified. But a person who assumes the title of physician or attorney without having earned it conventionally becomes an impostor. Secondly, the profession persuades the community to institute in its behalf a licensing system for screening those qualified to practice the professional skill. A *sine qua non* for the receipt of the license is, of course, a duly granted professional title. Another prerequisite may be an examination before a board of inquiry whose personnel have been drawn from the ranks of the profession. Police power enforces the licensing system; persons practicing the professional skill without a license are liable to punishment by public authority.[12]

---

10   Parsons, *op. cit.*

11   To set up and run a school for floral decorating requires no approval from the national florists' association, but no school of social work could operate long without approval of the Council on Social Work Education.

12   Many nonprofessional occupations have also succeeded in obtaining licensing legislation in their behalf. Witness the plumbers, radio operators, and barbers, to mention a few. However, the sanctions applied against a person practicing a nonprofessional occupation are·much less severe than is the case when a professional occupation is similarly involved.

Among the professional privileges, one of the most important is that of confidentiality. To facilitate efficient performance, the professional encourages the client to volunteer information he otherwise would not divulge. The community regards this as privileged communication, shared solely between client and professional, and protects the latter legally from encroachments upon such confidentiality. To be sure, only a select few of the professions, notably medicine and law, enjoy this immunity. Its very rarity makes it the ultimate in professionalization. Another one of the professional privileges is a relative immunity from community judgment on technical matters. Standards for professional performance are reached by consensus within the profession and are based on the existing body of theory. The lay community is presumed incapable of comprehending these standards and, hence, of using them to identify malpractice. It is generally conceded that a professional's performance can be evaluated only by his peers.

The powers and privileges described above constitute a monopoly granted by the community to the professional group. Therefore, when an occupation strives toward professional status, one of its prime objectives is to acquire this monopoly. But this is difficult to achieve, because counter forces within the community resist strongly the profession's claims to authority. Through its associations the profession wages an organized campaign to persuade the community that it will benefit greatly by granting the monopoly. Specifically the profession seeks to prove: that the performance of the occupational skill requires specialized education; that those who possess this education, in contrast to those·who do not, deliver a superior service; and that the human need being served is of sufficient social importance to justify the superior performance.

REGULATIVE CODE OF ETHICS

The monopoly enjoyed by a profession vis-à-vis clients and community is fraught with hazards. A monopoly can be abused; powers and privileges can be used to protect vested interests against the public weal.[13] The professional group could peg the price of its services at an unreasonably high level; it could restrict the numbers entering the occupation to create a scarcity of personnel; it could dilute the caliber of its performance without community awareness; and it could frustrate forces within the occupation pushing for socially beneficial changes in practices.[14] Were such abuses to become conspicuous, widespread, and permanent, the community would, of course, revoke the profession's monopoly. This extreme measure is normally unnecessary, because every profession has a built-in regulative code which compels ethical behavior on the part of its members.

The profession's ethical code is part formal and part informal. The formal is the written code to which the professional usually swears upon being admitted to practice; this is best exemplified by the Hippocratic Oath of the medical profession. The informal is the unwritten code, which nonetheless carries the weight of formal prescriptions. Through its ethical code the profession's commitment to the social welfare becomes a matter of public record; thereby insuring for itself the continued confidence of the community. Without such confidence the profession could not retain its monopoly. To be sure, self-regulative codes are characteristic of all occupations, nonprofessional as well as professional. However, a professional code is perhaps more explicit, systematic, and binding; it certainly possesses more altruistic overtones and is more public service-oriented.[15] These account for the frequent synonymous use of the terms "professional" and "ethical" when applied to occupational behavior.

While the specifics of their ethical codes vary among the professions, the essentials are uniform. These may be described in terms of client-professional and colleague-colleague relations.

Toward the client the professional must assume an emotional neutrality. He must provide service to

13 Abraham Flexner, "Is Social Work a Profession?" in *Proceedings of the National Conference of Charities and Corrections* (Chicago: 1915), pp. 576-590.

Robert K. Merton, "Bureaucratic Structure and Personality," in Alvin Gouldner (ed.), *Studies in Leadership* (New York: Harper &: Row, Publishers, 1950), pp. 67-79.

14 Merton, *op. cit.*

15 Flexner, *op. cit.* Parsons, *op. cit.*

whoever requests it, irrespective of the requesting client's age, income, kinship, politics, race, religion, sex, and social status. A nonprofessional may withhold his services on such grounds without, or with minor, censure; a professional cannot. Parsons calls this element in professional conduct *universalism*. In other words, only in his extraoccupational contacts can the professional relate to others on particularistic terms, i.e., as particular individuals with concrete personalities attractive or unattractive to him. In his client contacts particularistic considerations are out of place. Parsons also calls attention to the element of *disinterestedness* in the professional-client relationship.[16] In contrast to the nonprofessional, the professional is motivated less by self-interest and more by the impulse to perform maximally. The behavior corollaries of this service orientation are many. For one, the professional must, under all circumstances, give maximum caliber service. The nonprofessional can dilute the quality of his commodity or service to fit the size of the client's fee; not so the professional. Again, the professional must be prepared to render his services upon request, even at the sacrifice of personal convenience.

The ethics governing colleague relationships demand behavior that is cooperative, equalitarian, and supportive. Members of a profession share technical knowledge with each other. Any advance in theory and practice made by one professional is quickly disseminated to colleagues through the professional associations.[17] The proprietary and quasi-secretive attitudes toward discovery and invention prevalent in the industrial and commercial world are out of place in the professional. Also out of place is the blatant competition for clients which is the norm in so many nonprofessional pursuits. This is not to gainsay the existence of intraprofessional competition; but it is a highly regulated competition, diluted with cooperative ingredients which impart to it its characteristically restrained quality. Colleague relations must be equalitarian; intraprofessional recognition should ideally be based solely upon performance in

practice and/or contribution to theory.[18] Here, too, particularistic considerations must not be allowed to operate. Finally, professional colleagues must support each other vis-à-vis clientele and community. The professional must refrain from acts which jeopardize the authority of colleagues and must sustain those whose authority is threatened.[19]

The ways and means whereby a profession enforces the observance of its ethical code constitute a case study in social control. Self-discipline is achieved informally and formally.

Informal discipline consists of the subtle and the not-so-subtle pressures that colleagues exert upon one another. An example in this connection is the phenomenon of consultation and referral.[20] Consultation is the practice of inviting a colleague to participate in the appraisal of the client's need and/or in the planning of the service to be rendered. Referral is the practice of affording colleagues access to a client or an appointment. Thus, one colleague may refer his client to another, because lack of time or skill prevents his rendering the needed service; or he may recommend another for appointment by a prospective employer. Since professional ethics precludes aggressive competition and advertising, consultation and referral constitute the principal source of work to a professional. The consultation-referral custom involves professional colleagues in a system of reciprocity which fosters mutual interdependence. Interdependence facilitates social control; chronic violation of professional etiquette arouses colleague resentment, resulting in the cessation of consultation requests and referrals.

A more formal discipline is exercised by the professional associations, which possess the power to criticize or to censure, and in extreme cases to bar recalcitrants. Since membership in good standing in the professional associations is a *sine qua non* of professional success, the prospect of formal disciplinary action operates as a potent force toward conformity.

---

16 Parsons, *op. cit.*

17 Arlien Johnson, "Professional Standards and How They Are Attained," *Journal of American Dental Association,* 31 (September 1944), 1181-1189.

18 Flexner, *op. cit.*

19 This partly explains why physicians do not testify against each other in malpractice suits.

20 Hall, *op. cit.*

THE PROFESSIONAL CULTURE

Every profession operates through a network of formal and informal groups. Among the formal groups, first there are the organizations through which the profession performs its services, these provide the institutionalized setting where professional and client meet. Examples of such organizations are hospital, clinic, university, law office, engineering firm, or social agency. Secondly, there are the organizations whose functions are to replenish the profession's supply of talent and to expand its fund of knowledge. These include the educational and the research centers. Third among the formal groups are the organizations which emerge as an expression of the growing consciousness-ofkind on the part of the profession's members, and which promote so-called group interests and aims. These are the professional associations. Within and around these formal organizations extends a filigree of informal groupings: the multitude of small, closely knit clusters of colleagues. Membership in these cliques is based on a variety of affinities: specialties within the profession; affiliations with select professional societies; residential and work propinquity; family, religious, or ethnic background; and personality attraction.

The interaction of social roles required by these formal and informal groups generate a social configuration unique to the profession, *viz.*, a professional culture. All occupations are characterized by formal and informal groupings; in this respect the professions are not unique. What is unique is the culture thus begotten. If one were to single out the attribute that most effectively differentiates the professions from other occupations, this is it. Thus we can talk of a professional culture as distinct from a nonprofessional culture. Within the professions as a logical class each profession develops its own subculture, a variant of the professional culture; the engineering subculture, for example, differs from the subcultures of medicine and social work. In the subsequent discussion, however, we will treat the culture of the professions as a generic phenomenon. The culture of a profession consists of its *values, norms*, and *symbols*.

The social values of a professional group are its basic and fundamental beliefs, the unquestioned premises upon which its very existence rests. Foremost among these values is the essential worth of the service which the professional group extends to the community. The-profession considers that the service is a social good and that community welfare would be immeasurably impaired by its absence. The twin concepts of professional authority and monopoly also possess the force of a group value. Thus, the proposition that in all service-related matters the professional group is infinitely wiser than the laity is regarded as beyond argument. Likewise nonarguable is the proposition that acquisition by the professional group of a service monopoly would inevitably produce social progress. And then there is the value of rationality; that is, the commitment to objectivity in the realm of theory and technique. By virtue of this orientation, nothing of a theoretical or technical nature is regarded as sacred and unchallengeable simply because it has a history of acceptance and use.

The norms of a professional group are the guides to behavior in social situations. Every profession develops an elaborate system of these role definitions. There is a range of appropriate behaviors for seeking admittance into the profession, for gaining entry into its formal and informal groups, and for progressing within the occupation's hierarchy. There are appropriate modes of securing appointments, of conducting referrals, and of handling consultation. There are proper ways of acquiring clients, of receiving and dismissing them, of questioning and treating them, of accepting and rejecting them. There are correct ways of grooming a protégé, of recompensing a sponsor, and of relating to peers, superiors, or subordinates. There are even group-approved ways of challenging an outmoded theory, of introducing a new technique, and of conducting an intraprofessional controversy. In short, there is a behavior norm covering every standard interpersonal situation likely to recur in professional life.

The symbols of a profession are its meaning-laden items. These may include such things as: its insignias, emblems, and distinctive dress; its history, folklore, and argot; its heroes and its villains; and its stereotypes of the professional, the client, and the layman.

Comparatively clear and controlling group values, behavior norms, and symbols, which characterize the professions, are not to be encountered in nonprofessional occupations.

Our discussion of the professional culture would be incomplete without brief mention of one of its central concepts, the *career* concept. The term career is, as a rule, employed only in reference to a professional occupation. Thus, we do not talk about the career of a bricklayer or of a mechanic; but we do talk about the career of an architect or of a clergyman. At the heart of the career concept is a certain attitude toward work which is peculiarly professional. A career is essentially a *calling*, a life devoted to "good works."[21] Professional work is never viewed solely as a means to an end; it is the end itself. Curing the ill, educating the young, advancing science are values in themselves. The professional performs his services primarily for the psychic satisfactions and secondarily for the monetary compensations.[22] Self-seeking motives feature minimally in the choice of a profession; of maximal importance is affinity for the work. It is this devotion to the work itself which imparts to professional activity the service orientation and the element of disinterestedness. Furthermore, the absorption in the work is not partial, but complete; it results in a total personal involvement. The work life invades the after-work life, and the sharp demarcation between the work hours and the leisure hours disappears. To the professional person his work becomes his life.[23]

Hence the act of embarking upon a professional career is similar in some respects to entering a religious order. The same cannot be said of a nonprofessional occupation.

To succeed in his chosen profession, the neophyte must make an effective adjustment to the professional culture.[24] Mastery of the underlying body of theory and acquisition of the technical skills are in themselves insufficient guarantees of professional success. The recruit must also become familiar with and learn to weave his way through the labyrinth of the professional culture. Therefore, the transformation of a neophyte into a professional is essentially an acculturation process wherein he internalizes the social values, the behavior norms, and the symbols of the occupational group.[25] In its frustrations and rewards it is fundamentally no different from the acculturation of an immigrant to a relatively strange culture. Every profession entertains a stereotype of the ideal colleague; and, of course, it is always one who is thoroughly adjusted to the professional culture.[26] The poorly acculturated colleague is a deviant; he is regarded as "peculiar," "unorthodox," "annoying," and in extreme cases a "troublemaker." Whereas the professional group encourages innovation in theory and technique, it tends to discourage deviation from its social values and norms. In this internal contradiction, however, the professional culture is no different from the larger culture of society.

One of the principal functions of the professional schools is to identify and screen individuals who are prospective deviants from the professional culture. That is why the admission of candidates to professional education must be judged on grounds in

---

21 The term *calling* literally means a divine summons to undertake a course of action. Originally, it was employed to refer to religious activity. The Protestant Reformation widened its meaning to include economic activity as well. Henceforth divinely inspired "good works" were to be both secular and sacred in nature. Presumably, then, any occupational choice may be a response to divine summons. In this connection it is interesting to note that the German word for vocation is *Beruf,* a noun derived from the verb *Berufen,* to call.

22 Johnson, *op. cit.*

23 The all-pervading influence of work upon the lives of professionals results in interesting by-products. The members of a profession tend to associate with one another outside the work setting (Oswald Hall, "The Stages of a Medical Career," *op. cit.*). Their families mingle socially; leisure time is spent together; "shop talk" permeates social discourse; and a consensus develops. The profession thus becomes a whole social environment, nurturing characteristic social and political attitudes, patterns of consumption and recreation, and decorum and *Weltanschauung*. See Caplow, *op. cit.;* and William H. Form, "Toward an Occupational Social

Psychology," *Journal of Social Psychology,* 24 (February 1946), 85-99.

24 Oswald Hall, "The Stages of a Medical Career" and "Types of Medical Careers," *op. cit.*

25 R. Clyde White, "'Social Workers in Society': Some Further Evidence," *Social Work Journal,* 34 (October 1953), 161-164.

26 The laity also entertain a stereotypic image of the professional group. Needless to say, the layman's conception and the professional's self-conception diverge widely, because they are fabricated out of very different experiences. The layman's stereotype is frequently a distortion of reality, being either an idealization or a caricature of the professional type.

addition to and other than their academic qualifications.[27] Psychic factors presaging favorable adjustment to the professional culture are granted an importance equivalent to mental abilities. The professional school provides test situations through initial and graduated exposures of the novice to the professional culture. By his behavior in these social situations involving colleagues, clients, and community, the potential deviant soon reveals himself and is immediately weeded out. Comparable preoccupation with the psychic prerequisites of occupational adjustment is not characteristic of nonprofessional occupations.

IMPLICATIONS FOR SOCIAL WORK

The picture of the professions just unveiled is an ideal type. In the construction of an ideal type some exaggeration of reality is unavoidable, since the intent is to achieve an internally coherent picture. One function of the ideal type is to structure reality in such manner that discrete, disparate, and dissimilar phenomena become organized, thereby bringing order out of apparent disorder. We now possess a model of a profession that is much sharper and clearer than the actuality that confronts us when we observe the occupational scene. What is the utility of this model for social work?

The preoccupation of social workers with professionalization has been a characteristic feature of the social work scene for years. Flexner,[28] Johnson,[29] Hollis and Taylor,[30] and others have written on the subject, proposing criteria which must be met if social work is to acquire professional status. Whenever social workers convene, there is the constant reaffirmation of the urgency to achieve the recognition from the community befitting a profession. The union of the seven separate organizations into the National Association of Social Workers is generally regarded as an important milestone in social work history, precisely because of its potential stimulus toward professionalization.

In view of all this, it is proper for social workers to possess clear conceptions of that which they so fervently seek. The model of the professions portrayed above should contribute to such clarification; it should illuminate the goal for which social workers are striving. It is often contended that social work is still far from having attained professional status.[31] But this is a misconception. When we hold up social work against the model of the professions presented above, it does not take long to decide whether to classify it within the professional or the nonprofessional occupations. Social work is already a profession; it has too many points of congruence with the model to be classifiable otherwise. Social work is, however, seeking to rise within the professional hierarchy, so that it, too, might enjoy maximum prestige, authority, and monopoly which presently belong to a few top professions.

The model presented above should also serve to sensitize social workers to anticipate some of the problems that continued professionalization must inevitably precipitate. The model indicates that progressive professionalization will involve social workers in novel relationships with clients, colleagues, agency, community, and other professions. In concluding this paper we refer briefly to one such problem. It is no secret that social workers are not all uniformly enthusiastic about the professionalization of social work. Bisno[32] has given verbalization to a prevailing apprehension that social workers might have to scuttle their social-action heritage as a price of achieving the public acceptance accorded a profession. Extrapolation from the sociologists' model of the professions suggests a reality basis for these fears. It suggests that the attainment of professional prestige, authority, and monopoly by social workers will undoubtedly carry disturbing implications for the social action and social reform components of social work philosophy. The anticipated developments will compel social workers to rethink and redefine the societal role of their profession.

27 Oswald Hall, "Sociological Research in the Field of Medicine: Progress and Prospects," *op. cit.*

28 Flexner, *op. cit.*

29 Johnson, *op. cit.*

30 Ernest V. Hollis and Alice L. Taylor, *Social Work Education in the United States* (New York: Columbia University Press, 1951).

31 Flexner considered that the social work of his day was not a profession. Hollis and Taylor regard present-day social work as still in its early adolescence.

32 Herbert Bisno, "How Social Will Social Work Be?" *Social Work,* 1, No. 2 (April 1956), 12-18.

These and other dilemmas flowing from professionalization are bound to tax the best minds among
social workers for their resolution. In this connection
a proper understanding of the attributes of a profession would seem to be indispensable.

Code of Ethics for Engineers

National Society of Professional Engineers, January 2006

PREAMBLE

Engineering is an important and learned profession. As members of this profession, engineers are expected to exhibit the highest standards of honesty and integrity. Engineering has a direct and vital impact on the quality of life for all people. Accordingly, the services provided by engineers require honesty, impartiality, fairness, and equity, and must be dedicated to the protection of the public health, safety, and welfare. Engineers must perform under a standard of professional behavior that requires adherence to the highest principles of ethical conduct.

I. FUNDAMENTAL CANONS

Engineers, in the fulfillment of their professional duties, shall:

1. Hold paramount the safety, health, and welfare of the public.
2. Perform services only in areas of their competence.
3. Issue public statements only in an objective and truthful manner.
4. Act for each employer or client as faithful agents or trustees.
5. Avoid deceptive acts.
6. Conduct themselves honorably, responsibly, ethically, and lawfully so as to enhance the honor, reputation, and usefulness of the profession.

II. RULES OF PRACTICE

1. Engineers shall hold paramount the safety, health, and welfare of the public.
   a. If engineers' judgment is overruled under circumstances that endanger life or property, they shall notify their employer or client and such other authority as may be appropriate.
   b. Engineers shall approve only those engineering documents that are in conformity with applicable standards.
   c. Engineers shall not reveal facts, data, or information without the prior consent of the client or employer except as authorized or required by law or this Code.
   d. Engineers shall not permit the use of their name or associate in business ventures with any person or firm that they believe is engaged in fraudulent or dishonest enterprise.
   e. Engineers shall not aid or abet the unlawful practice of engineering by a person or firm.
   f. Engineers having knowledge of any alleged violation of this Code shall report thereon to appropriate professional bodies and, when relevant, also to public authorities, and cooperate with the proper authorities in furnishing such information or assistance as may be required.

2. Engineers shall perform services only in the areas of their competence.

   a. Engineers shall undertake assignments only when qualified by education or experience in the specific technical fields involved.

   b. Engineers shall not affix their signatures to any plans or documents dealing with subject matter in which they lack competence, nor to any plan or document not prepared under their direction and control.

   c. Engineers may accept assignments and assume responsibility for coordination of an entire project and sign and seal the engineering documents for the entire project, provided that each technical segment is signed and sealed only by the qualified engineers who prepared the segment.

3. Engineers shall issue public statements only in an objective and truthful manner.

   a. Engineers shall be objective and truthful in professional reports, statements, or testimony. They shall include all relevant and pertinent information in such reports, statements, or testimony, which should bear the date indicating when it was current.

   b. Engineers may express publicly technical opinions that are founded upon knowledge of the facts and competence in the subject matter.

   c. Engineers shall issue no statements, criticisms, or arguments on technical matters that are inspired or paid for by interested parties, unless they have prefaced their comments by explicitly identifying the interested parties on whose behalf they are speaking, and by revealing the existence of any interest the engineers may have in the matters.

4. Engineers shall act for each employer or client as faithful agents or trustees.

   a. Engineers shall disclose all known or potential conflicts of interest that could influence or appear to influence their judgment or the quality of their services.

   b. Engineers shall not accept compensation, financial or otherwise, from more than one party for services on the same project, or for services pertaining to the same project, unless the circumstances are fully disclosed and agreed to by all interested parties.

   c. Engineers shall not solicit or accept financial or other valuable consideration, directly or indirectly, from outside agents in connection with the work for which they are responsible.

   d. Engineers in public service as members, advisors, or employees of a governmental or quasi-governmental body or department shall not participate in decisions with respect to services solicited or provided by them or their organizations in private or public engineering practice.

   e. Engineers shall not solicit or accept a contract from a governmental body on which a principal or officer of their organization serves as a member.

5. Engineers shall avoid deceptive acts.

   a. Engineers shall not falsify their qualifications or permit misrepresentation of their or their associates' qualifications. They shall not misrepresent or exaggerate their responsibility in or for the subject matter of prior assignments. Brochures or other presentations incident to the solicitation of employment shall not misrepresent pertinent facts concerning employers, employees, associates, joint venturers, or past accomplishments.

   b. Engineers shall not offer, give, solicit, or receive, either directly or indirectly, any contribution to influence the award of a contract by public authority, or which may be reasonably construed by the public as having the effect or intent of influencing the awarding of a contract. They shall not offer any gift or other valuable consideration in order to secure work. They shall not pay a commission, percentage, or brokerage fee in order to secure work, except to a bona fide employee or bona fide established commercial or marketing agencies retained by them.

III. PROFESSIONAL OBLIGATIONS

1. Engineers shall be guided in all their relations by the highest standards of honesty and integrity.

    a. Engineers shall acknowledge their errors and shall not distort or alter the facts.

    b. Engineers shall advise their clients or employers when they believe a project will not be successful.

    c. Engineers shall not accept outside employment to the detriment of their regular work or interest. Before accepting any outside engineering employment, they will notify their employers.

    d. Engineers shall not attempt to attract an engineer from another employer by false or misleading pretenses.

    e. Engineers shall not promote their own interest at the expense of the dignity and integrity of the profession.

2. Engineers shall at all times strive to serve the public interest.

    a. Engineers shall seek opportunities to participate in civic affairs; career guidance for youths; and work for the advancement of the safety, health, and well-being of their community.

    b. Engineers shall not complete, sign, or seal plans and/or specifications that are not in conformity with applicable engineering standards. If the client or employer insists on such unprofessional conduct, they shall notify the proper authorities and withdraw from further service on the project.

    c. Engineers shall endeavor to extend public knowledge and appreciation of engineering and its achievements.

    d. Engineers shall strive to adhere to the principles of sustainable development[1] in order to protect the environment for future generations.

3. Engineers shall avoid all conduct or practice that deceives the public.

    a. Engineers shall avoid the use of statements containing a material misrepresentation of fact or omitting a material fact.

    b. Consistent with the foregoing, engineers may advertise for recruitment of personnel.

    c. Consistent with the foregoing, engineers may prepare articles for the lay or technical press, but such articles shall not imply credit to the author for work performed by others.

4. Engineers shall not disclose, without consent, confidential information concerning the business affairs or technical processes of any present or former client or employer, or public body on which they serve.

    a. Engineers shall not, without the consent of all interested parties, promote or arrange for new employment or practice in connection with a specific project for which the engineer has gained particular and specialized knowledge.

    b. Engineers shall not, without the consent of all interested parties, participate in or represent an adversary interest in connection with a specific project or proceeding in which the engineer has gained particular specialized knowledge on behalf of a former client or employer.

5. Engineers shall not be influenced in their professional duties by conflicting interests.

    a. Engineers shall not accept financial or other considerations, including free engineering designs, from material or equipment suppliers for specifying their product.

    b. Engineers shall not accept commissions or allowances, directly or indirectly, from contractors or other parties dealing with clients or employers of the engineer in connection with work for which the engineer is responsible.

6. Engineers shall not attempt to obtain employment or advancement or professional engagements by untruthfully criticizing other engineers, or by other improper or questionable methods.

    a. Engineers shall not request, propose, or accept a commission on a contingent basis under circumstances in which their judgment may be compromised.

b. Engineers in salaried positions shall accept part-time engineering work only to the extent consistent with policies of the employer and in accordance with ethical considerations.

c. Engineers shall not, without consent, use equipment, supplies, laboratory, or office facilities of an employer to carry on outside private practice.

7. Engineers shall not attempt to injure, maliciously or falsely, directly or indirectly, the professional reputation, prospects, practice, or employment of other engineers. Engineers who believe others are guilty of unethical or illegal practice shall present such information to the proper authority for action.

a. Engineers in private practice shall not review the work of another engineer for the same client, except with the knowledge of such engineer, or unless the connection of such engineer with the work has been terminated.

b. Engineers in governmental, industrial, or educational employ are entitled to review and evaluate the work of other engineers when so required by their employment duties.

c. Engineers in sales or industrial employ are entitled to make engineering comparisons of represented products with products of other suppliers.

8. Engineers shall accept personal responsibility for their professional activities, provided, however, that engineers may seek indemnification for services arising out of their practice for other than gross negligence, where the engineer's interests cannot otherwise be protected.

a. Engineers shall conform with state registration laws in the practice of engineering.

b. Engineers shall not use association with a nonengineer, a corporation, or partnership as a "cloak" for unethical acts.

9. Engineers shall give credit for engineering work to those to whom credit is due, and will recognize the proprietary interests of others.

a. Engineers shall, whenever possible, name the person or persons who may be individually responsible for designs, inventions, writings, or other accomplishments.

b. Engineers using designs supplied by a client recognize that the designs remain the property of the client and may not be duplicated by the engineer for others without express permission.

c. Engineers, before undertaking work for others in connection with which the engineer may make improvements, plans, designs, inventions, or other records that may justify copyrights or patents, should enter into a positive agreement regarding ownership.

d. Engineers' designs, data, records, and notes referring exclusively to an employer's work are the employer's property. The employer should indemnify the engineer for use of the information for any purpose other than the original purpose.

e. Engineers shall continue their professional development throughout their careers and should keep current in their specialty fields by engaging in professional practice, participating in continuing education courses, reading in the technical literature, and attending professional meetings and seminars.

Footnote 1 "Sustainable development" is the challenge of meeting human needs for natural resources, industrial products, energy, food, transportation, shelter, and effective waste management while conserving and protecting environmental quality and the natural resource base essential for future development.

AS REVISED JANUARY 2006

"By order of the United States District Court for the District of Columbia, former Section 11(c) of the NSPE Code of Ethics prohibiting competitive bidding, and all policy statements, opinions, rulings or other guidelines interpreting its scope, have been rescinded as unlawfully interfering with the legal right of engineers, protected under the antitrust laws, to provide price information to prospective clients; accordingly, nothing contained in the NSPE Code of Ethics, policy statements, opinions, rulings or other guidelines prohibits the submission of price quotations or competitive bids for engineering services at any time or in any amount."

Statement by NSPE Executive Committee

In order to correct misunderstandings which have been indicated in some instances since the issuance of the Supreme Court decision and the entry of the Final Judgment, it is noted that in its decision of April 25, 1978, the Supreme Court of the United States declared: "The Sherman Act does not require competitive bidding."

It is further noted that as made clear in the Supreme Court decision:

1. Engineers and firms may individually refuse to bid for engineering services.
2. Clients are not required to seek bids for engineering services.
3. Federal, state, and local laws governing procedures to procure engineering services are not affected, and remain in full force and effect.
4. State societies and local chapters are free to actively and aggressively seek legislation for professional selection and negotiation procedures by public agencies.
5. State registration board rules of professional conduct, including rules prohibiting competitive bidding for engineering services, are not affected and remain in full force and effect. State registration boards with authority to adopt rules of professional conduct may adopt rules governing procedures to obtain engineering services.
6. As noted by the Supreme Court, "nothing in the judgment prevents NSPE and its members from attempting to influence governmental action ..."

NOTE: In regard to the question of application of the Code to corporations vis-a-vis real persons, business form or type should not negate nor influence conformance of individuals to the Code. The Code deals with professional services, which services must be performed by real persons. Real persons in turn establish and implement policies within business structures. The Code is clearly written to apply to the Engineer, and it is incumbent on members of NSPE to endeavor to live up to its provisions. This applies to all pertinent sections of the Code.

National Society of Professional Engineers®

1420 King Street
Alexandria, Virginia 22314-2794
Phone: 703/684-2800   Fax: 703/836-4875
www.nspe.org
Publication date as revised: January 2006
Publication #1102

Risk at the Roots

By Peter L. Bernstein

The dominance of rational models of human behaviour has led to the notion that the quantification of variables can guarantee safety against risk. Drawing inspiration from Pascal, Bernoulli and Gould, Peter Bernstein argues that risk is not only inevitable but valuable and that a predictable world would be the most dangerous of all

BUSINESS MANAGEMENT is a process rooted in the future. We do not know the future. Then how in the world can we make decisions? In this article I look at this question from the viewpoint of two great philosophers of the past and one distinguished scientist of the present. The bottom line is that the consequences of the choices we face should outweigh probabilities of the outcomes we expect, but humans do have free will and can make rational choices even when the future is uncertain.

The first approach in both historical order and order of importance dates back nearly 350 years. My source is Blaise Pascal, a Parisian who was one of the most famous mathematicians and philosophers of his time. It was he, together with the lawyer and mathematical genius Robert de Fermat, who first propounded the principles of probability in 1654. Pascal spent most of his life alternating between a life of casinos and fleshpots, followed by intense religious periods marked by total abstinence from these enjoyments. In the end, the religious mode won: Pascal retired to the monastery of Port-Royal—which was supported by the earnings of the first commercial bus company in Paris, of which Pascal was the founder.

Scribbled in the manuscript of his autobiography, Pascal set forth what has come to be known as le pari de Pascal, or Pascal's wager. Pascal began by daring to ask: 'God is, or he is not. Which way should we incline? Reason cannot answer.'

We have to reconstitute the question before we can answer it. Is an outcome in which 'God is' preferable—more valuable in some sense—to an outcome in which 'God is not?' The issue is not belief in God, for that cannot be reduced to a decision in which you wake up one day and say, 'Today I have decided to believe (or not believe) in God.' Nor do our guesses as to the probability that 'God is' help us very much since we have no reliable method for testing or falsifying the hypothesis.

To Hedge or Not?

Suppose you act as though God is and lead a life of virtue, unselfishness, and abstinence. Suppose you are wrong and God is not. You will have passed up some fun, but there would also be rewards. Now suppose you act as though God is not, hanging out in the casinos and fleshpots and placing 'me' above all else. And suppose you are wrong on this side of the bet. You may have had a lot of goodies in your lifetime, but you will suffer damnation into eternity. The value of the bet that God is, is infinitely greater than the value of the bet that God is not. The probability that God is or God is not is irrelevant.

Now translate this line of analysis into business decisions. Suppose you are planning to launch a new product into another country where the market for this kind of product is growing vigorously. The probabilities of success appear high, but this market is also highly competitive. Differentiating your brand is difficult; price is the dominant variable. On the other hand, given the interest rate spreads and the exchange market's optimistic outlook for this nation's currency, you see no need to hedge the foreign currency bet at this time. Should you hedge or not hedge?

Suppose you decide to hedge and are wrong: exchange rates turn out to be stable. Although you will have incurred the cost of hedging, you will still be able to price your product correctly for this competitive market, no matter what happens to the exchange rate. Now suppose you agree with the forecast of the foreign exchange market and refuse to hedge but the currency surprises the market by depreciating. You will receive less sterling than before for each unit you sell abroad in a market where raising your local currency price even by a little will put you out of business.

You have bet that God is not, but God is. You cannot afford to make that bet. Regardless of what you believe the currency is going to do, regardless of the probabilities, maintaining a competitive local price for your product is the single most important element in success. Hence, you must hedge or not take the risk of entering that market.

Philosophers on Risk

Fast forward about 50 years to a scientist and mathematician named Bernoulli, a member of a family that was extraordinarily talented but also nasty and mean. In 1703, Jacob Bernoulli wrote a letter to Gottfried von Leibniz in which he observed that: 'It is strange that we know the odds of throwing a seven instead of an eight with a pair of dice, but we do not know the probability that a man of 20 will outlive a man of 60.' Jacob proposed an experiment in which he would compare a large number of pairs of men of various ages to see if he could deduce the probabilities from that evidence.

Leibniz took a dim view of the suggestion. 'Nature', he wrote, 'has established patterns originating in the turn of events, but only for the most part.' In those

days, correspondence of this type was always carried on in Latin, but Leibniz placed such importance on those last five critical words that he set them in Greek. Never forget them.

His message rings true across the centuries. You cannot escape uncertainty. No mathematical model works to perfection. Statisticians are satisfied when they can demonstrate that a model works with only a 5% probability that its results are due to chance, but that still leaves 5% that we do not understand, cannot model, and can cause all kinds of mischief if we mechanically make decisions based on the model. No event is without cause, so ascribing a probability to chance, or luck, is merely assigning a number to our ignorance.

As patterns repeat themselves only for the most part, we must use the tools of risk management—the art of survival when our forecasts of the unknown future turn out to be wrong. In the example above, the currency did depreciate, even though the probabilities were against it. The other part matters every bit as much as the most part.

Now I turn to the brilliant American evolutionary biologist, Stephen Jay Gould. In his recent book, Full House: The Spread of Excellence from Plato to Darwin, Gould writes about the moment some 15 years ago, when the doctors informed him that he had developed a deadly form of cancer called mesothelioma. The median life expectancy for this disease, they reported, was eight months. How does one react to a diagnosis like that? Give way to devastation? Carry on as though the problem did not exist? Or pray?

In a manner of speaking, Gould chose prayer. His religion is Darwinian evolution, a theoretical structure that describes a reality in which causes have effects but the variety of possible effects from any given cause is virtually limitless. Consequently, the effects are not predictable. In this religion, as Gould describes it, 'variation stands as the fundamental reality and calculated averages become abstractions.'

Gould reasoned that it would be an error to view the forecast offered by his doctors as the most likely outcome for any single individual. 'I am not a measure of central tendency, either mean or median. I am one single human being with mesothelioma, and I want the best assessment of my own chances.' His crucial insight was to recognise that there is little room between the absolute minimum of zero life expectancy (dropping dead at the moment of diagnosis) and the median value of eight months: 'Half of the variation must be scrunched up into this left half of the curve between the minimum and the median. But the right half, in principle, extends out forever, or at least into extreme old age.

From that deduction stemmed the good news. Gould noted that many factors favoured his potential location in the right tail—he was young and full of fight, was blessed with a supportive family, lived in a city that offered outstanding medical treatment, and was fortunate in having his illness diagnosed early on. He fought and won. If he had been mesmerised by the central tendency—by the median—he might well have ended up as a suicide. Yet, 15 years later he is a healthy and vigorous man.

Gould goes on to make the point in a general sense: 'We are still suffering from a legacy as old as Plato … a tendency to abstract a single ideal or average as the "essence" of a system and to devalue or ignore variation among the individuals that constitute the full population … We have never put aside this distinctive view that populations of actual individuals form a set of accidents, a collection of flawed examples, each necessarily incorrect and capable only of approaching the ideal to a certain extent …We regard variation as a pool of inconsequential happenstances, valuable largely because we can use the spread to calculate an average, which we may then regard as the best approach to an essence.'

In short, all of us depend far too much on measures of central tendency and therefore allow ourselves to be mesmerised by the hole rather than the doughnut. We base our decisions on trends, coefficients of correlation, normal distributions, even more arcane gadgets, and above all on the average.

But the sum and substance of risk management is in the recognition of variety. Losses stem, not from average results, but from deviations from the average or norm, from the outliers at the far reaches of the tails, and from outcomes never even imagined. Surprise pursues us relentlessly because we can never have all the information we need for a correct forecast every time. However, the incomplete information we do have is overwhelming, and getting more so

all the time. We probably could not handle all the information we need, even if it were available to us!

Risk Management

We have no choice but to simplify. In particular, we are addicted to smoothing the data, to removing the bumps and wrinkles in order to identify the essence, the trend, the values that we would like to believe are the true values. We prefer seasonally adjusted data to raw data, we find it easier to understand year-over-year rates of change than sequential monthly or quarterly changes, and we like moving averages and trendlines. The revered discounted cash flow model itself is a smoothing tool that glides us from the present into some distant point in the future.

Accrual accounting is an especially vivid form of smoothing, an act of faith that we know what the future holds. It records as revenues money not yet received, and that may not be received. It excludes from expenses money actually laid out on assets that are expected to produce cash revenues in the future, but they may fail to produce.

Hence, although every form of smoothing assumes that variations from a specified path are transitory or systematic, major risks lurk in the interstices of all those techniques. We are like people who are led to believe they are seeing moving pictures at the cinema when all they are seeing is a rapidly moving sequence of frames of still pictures. Simplification lures us into the trap that Gould warns we set for ourselves with our demand for the 'essence' in preference to the variation, for simplification is impossible without the averages and the other measures of central tendency.

We do concern ourselves with the big risks admittedly when changes are so great that the new environment seems to have had no past at all—events like the OPEC shock of 1973, the abrupt arrival of disinflation during 1982, the recent far-reaching transformation in the role of government, and the all-too-frequent powerful smashings of well-established parameters such as former highs and lows in interest rates, equity valuation, or yield spreads. These are the kinds of change where all the old rules go out the window and we have to learn an entirely new game.

We characterise these moments as discontinuities, paradigm shifts, or regime changes.

Gould and his fellow Darwinians would laugh at our nomenclature. They refuse to recognise discontinuities, paradigm shifts, or regime changes. Before discontinuity, we must have continuity; before a paradigm shift, we must have a paradigm; before a regime shift, we must have a regime. We must have trends, norms, averages, and other indicators of central tendency before we can have shocks to the system.

But the Darwinian would argue that diversity in nature is so great and every event could lead to such a great variety of outcomes, that there is no such thing as a state of nature that we can identify as a paradigm. The trick in risk management, perhaps, is in recognising that normal is not a state of nature but of transition, and trend is not destiny. Yet Gould is sending us a positive message. Variety is the spice of life. Just imagine what life would be like if everything closely resembled its measure of central tendency—if we were all clones of just one particular human being, or if all corporations were merged into just USA Inc. or, even worse, into Globalisation Inc.

A profound and overriding point is embedded in that suggestion: life would not only be intolerably boring without variety, but also intolerably risky! Variety is what makes risk management possible, for—by definition—diversity is the necessary condition for diversification and for hedging. It means that we have more than one basket in which to deposit the eggs we want to carry. As a result, society is willing to take on more risk than if we had just one basket. Indeed, variety is at the essence of survival. We do not all come down with mesothelioma at the same instant, all ships do not sink together, all stocks do not go up and down together, and different capital assets have different cyclical patterns.

The philosophical issue runs even deeper. Take away variety and nothing remains but averages with zero standard deviations and central tendencies always on the mark. Take away variety, in other words, and we eliminate uncertainty. And what is so good about that? Once you eliminate uncertainty, you have crushed human free will, for then everything is predetermined and decision-making becomes an obsolescent skill.

Rules of the Board for Professional Engineers and Land Surveyors

California Code of Regulations Title 16, Division 5 §§ 400–476

ARTICLE 1. GENERAL PROVISIONS

400. Introduction.

These regulations are adopted by the Board in order to implement and make specific the Professional Engineers' Act and the Land Surveyors' Act, Business and Professions Code Sections 6700, et seq., and 8700, et seq., respectively.

401. Declaratory Decisions.

No decision or opinion issued by or on behalf of the Board for Professional Engineers and Land Surveyors shall be considered a declaratory decision pursuant to Government Code section 11465.10 through 11465.70 unless the decision or opinion specifically states that it is a declaratory decision issued pursuant to Government Code sections 11465.10 through 11465.70.

403. Location of Offices.

The principal office of the Board is 2535 Capitol Oaks Drive, Suite 300, Sacramento, 95833-2944. All correspondence relating to the activities of the Board, including applications, renewals, and remittances, shall be directed to the Board's office.

404. Definitions.

For the purpose of the rules and regulations contained in this chapter, the following terms are defined.

No definition contained herein authorizes the practice of engineering as defined in the Professional Engineers Act.

(a) "Agricultural engineering" is that branch of professional engineering which requires such education and experience as is necessary to understand and apply engineering principles to the design, construction, and use of specialized equipment, machines structures and materials relating to the agricultural industry and economy. It requires knowledge of the engineering sciences relating to physical properties and biological variables of foods and fibers; atmospheric phenomena as they are related to agricultural operations; soil dynamics as related to traction, tillage and plant-soil-water relationships; and human factors relative to safe design and use of agricultural machines. The safe and proper application and use of agricultural chemicals and their effect on the environment are also concerns of the agricultural engineers. The above definition of agricultural engineering shall not be construed to permit the practice of civil, electrical or mechanical engineering, nor professional forestry.

(b) "Board" means the Board for Professional Engineers and Land Surveyors.

(c) "Chemical engineering" is that branch of professional engineering which embraces studies or activities relating to the development and application of processes in which chemical or physical changes of materials are involved. These processes are usually resolved into a coordinated series of unit physical operations and unit chemical processes. It is concerned with the research, design, production, operational, organizational, and economic aspects of the above. The above definition of chemical engineering shall not be construed to permit the practice of civil, electrical or mechanical engineering.

(d) "Civil engineer" refers to a person who holds a valid license in the branch of civil engineering, as defined in Section 6702 of the Code.

(e) "Civil engineering" is that branch of professional engineering as defined in Section 6731 of the Code.

(f) "Code" means the Business and Professions Code.

(g) "Consulting engineer" refers to any professional engineer who holds a valid license under the provisions of the code, or a person who possesses a valid authorization issued pursuant to Section 6732.2 of the Code, or a person who holds a valid exemption from provisions of the chapter as provided for in Sections 6704 and 6732.1 of the Code.

(h) "Control system engineering" is that branch of professional engineering which requires such education and experience as is necessary to understand the science of instrumentation and automatic control of dynamic processes; and requires the ability to apply this knowledge to the planning, development, operation, and evaluation of systems of control so as to insure the safety and practical operability of such processes. The above definition of control system engineering shall not be construed to permit the practice of civil, electrical, or mechanical engineering.

(i) "Corrosion engineering" is that branch of professional engineering which requires such education and experience as is necessary to understand the environmental corrosion behavior of materials; and requires the ability to apply this knowledge by recommending procedures for control, protection and cost effectiveness, resulting from the investigation of corrosion causes or theoretical reactions. The above definition of corrosion engineering shall not be construed to permit the practice of civil, electrical, or mechanical engineering.

(j) "Electrical engineer" refers to a person who holds a valid license in the branch of electrical engineering, as defined in Section 6702.1 of the Code.

(k) "Electrical engineering" is that branch of professional engineering as defined in Section 6731.5 of the Code.

(l) "Engineer-in-training" refers to a person who has been granted a certificate as an "engineer-in-training" in accordance with Section 6756 of the Code.

(m) "Fire protection engineering" is that branch of professional engineering which requires such education and experience as is necessary to understand the engineering problems relating to the safeguarding of life and property from fire and fire-related hazards; and requires the ability to apply this knowledge to the identification, evaluation, correction, or prevention of present or potential fire and fire related panic hazards in buildings, groups of buildings, or communities, and to recommend the arrangement and use of fire resistant building materials and fire detection and extinguishing systems, devices, and apparatus in order to protect life and property. The above definition of fire protection engineering shall not be construed to permit the practice of civil, electrical, or mechanical engineering.

(n) For the sole purpose of investigating complaints and making findings thereon under Sections 6775 and 8780 of the Code, "incompetence" as used in Sections 6775 and 8780 of the Code is defined as the lack of knowledge or ability in discharging professional obligations as a professional engineer or land surveyor.

(o) "Industrial engineering" is that branch of professional engineering which requires such education and experience as is necessary to investigate, to design, and to evaluate systems of persons, materials and facilities for the purpose of economical and efficient production, use, and distribution. It requires the application of specialized engineering knowledge of the mathematical and physical sciences, together with the principles and methods of engineering analysis and design to specify, predict, and to evaluate the results to be obtained from such systems. The above definition of industrial engineering shall not be construed to permit the practice of civil, electrical, or mechanical engineering.

(p) "Land surveying" is that practice defined in Section 8726 of the Code.

(q) "Land surveyor" refers to a person who holds a valid license as a land surveyor, as defined in Section 8701 of the Code.

(r) "Land surveyor-in-training" refers to a person who has been granted a certificate as a "land surveyor-in-training" in accordance with Section 8747 (a) of the Code.

(s) "Manufacturing engineering" is that branch of professional engineering which requires such education and experience as is necessary to understand and apply engineering procedures in manufacturing processes and methods of production of industrial commodities and products; and requires the ability to plan the practices of manufacturing, to research and develop the tools, processes, machines, and equipment, and to integrate the facilities and systems for producing quality products with optimal expenditure. The above definition of manufacturing engineering shall not be construed to permit the practice of civil, electrical, or mechanical engineering.

(t) "Mechanical engineer" refers to a person who holds a valid license in the branch of mechanical engineering, as defined in Section 6702.2 of the Code.

(u) "Mechanical engineering" is that branch of professional engineering as defined in Section 6731.6 of the Code.

(v) "Metallurgical engineering" is that branch of professional engineering, which requires such education and experience as is necessary to seek, understand and apply the principles of the properties and behavior of metals in solving engineering problems dealing with the research, development and application of metals and alloys; and the manufacturing practices of extracting, refining and processing of metals. The above definition of metallurgical engineering shall not be construed to permit the practice of civil, electrical, or mechanical engineering.

(w) For the sole purpose of investigating complaints and making findings thereon under Sections 6775 and 8780 of the Code, "negligence" as used in Sections 6775 and 8780 of the Code is defined as the failure of a licensee, in the practice of professional engineering or land surveying, to use the care ordinarily exercised in like cases by duly licensed professional engineers and land surveyors in good standing.

(x) "Nuclear engineering" is that branch of professional engineering which requires such education and experience as is necessary to apply the principles of nuclear physics to the engineering utilization of nuclear phenomena for the benefit of mankind; it is also concerned with the protection of the public from the potential hazards of radiation and radioactive materials. Nuclear engineering is primarily concerned with interaction of radiation and nuclear particles with matter. Nuclear engineering requires the application of specialized knowledge of the mathematical and physical sciences, together with the principles and methods of engineering design and nuclear analysis to specify, predict and evaluate the behavior of systems involving nuclear reactions, and to ensure the safe, efficient operation of these systems, their nuclear products and by-products. Nuclear engineering encompasses, but is not limited to, the planning and design of the specialized equipment and process systems of nuclear reactor facilities; and the protection of the public from any hazardous radiation produced in the entire nuclear reaction process. These activities include all aspects of the manufacture, transportation and use of radioactive materials. The above definition of nuclear engineering shall not be construed to permit the practice of civil, electrical, or mechanical engineering.

(y) "Petroleum engineering" is that branch of professional engineering which embraces studies or activities relating to the exploration, exploitation, location, and recovery of natural fluid hydrocarbons. It is concerned with research, design, production, and operation of devices, and the economic aspects of the above. The above definition of petroleum engineering shall not be construed to permit the practice of civil, electrical, or mechanical engineering.

(z) "Professional engineer" refers to a person engaged in the practice of professional engineering as defined in Section 6701 of the Code.

(aa) "Professional engineering" within the meaning of this chapter comprises the following branches: agricultural engineering, chemical engineering, civil engineering, control system engineering, corrosion engineering, electrical engineering, fire protection engineering, industrial engineering, manufacturing engineering, mechanical engineering, metallurgical engineering, nuclear engineering, petroleum engineering, quality engineering, safety engineering, and traffic engineering.

(bb) "Quality engineering" is that branch of professional engineering which requires such education and experience as is necessary to understand and apply the principles of product and service quality evaluation and control in the planning, development and operation of quality control systems, and the application and analysis of testing and inspection procedures; and requires the ability to apply metrology and statistical methods to diagnose and correct improper quality control practices to assure product and service reliability and conformity to prescribed standards. The above definition of quality engineering shall not be construed to permit the practice of civil, electrical, or mechanical engineering.

(cc) "Safety engineering" is that branch of professional engineering which requires such education and experience as is necessary to understand the engineering principles essential to the identification, elimination and control of hazards to people and property; and requires the ability to apply this knowledge to the development, analysis, production, construction, testing, and utilization of systems, products, procedures and standards in order to eliminate or optimally control hazards. The above definition of safety engineering shall not be construed to permit the practice of civil, electrical, or mechanical engineering.

(dd) "Soil engineer" refers to a civil engineer who holds a valid authorization to use the title "soil engineer," as provided in Section 6736.1 of the Code.

(ee) "Soil engineering," as it relates to the authorization to use the title "soil engineer," is the investigation and engineering evaluation of earth materials including soil, rock, groundwater and man-made materials and their interaction with earth retention systems, structural foundations and other civil engineering works. The practice involves application of the principles of soil mechanics and the earth sciences, and requires a knowledge of engineering laws, formulas, construction techniques and performance evaluation of civil engineering works influenced by earth materials.

The terms "geotechnical engineer" and "soils engineer" are deemed to be synonymous with the term "soil engineer."

(ff) "Structural engineer" refers to a civil engineer who holds a valid authorization to use the title "structural engineer," as provided in Section 6736 of the Code.

(gg) "Structural engineering" for the purposes of structural authority is the application of specialized civil engineering knowledge and experience to the design and analysis of buildings (or other structures) which are constructed or rehabilitated to resist forces induced by vertical and horizontal loads of a static and dynamic nature. This specialized knowledge includes familiarity with scientific and mathematical principles, experimental research data and practical construction methods and processes. The design and analysis shall include consideration of stability, deflection, stiffness and other structural phenomena that affect the behavior of the building (or other structure).

(hh) "Traffic engineering" is that branch of professional engineering which requires such education and experience as is necessary to understand the science of measuring traffic and travel and the human factors relating to traffic generation and flow; and requires the ability to apply this knowledge to planning, operating, and evaluating streets and highways and their networks, abutting lands and interrelationships with other modes of travel, to provide safe and efficient movement of people and goods. The above definition of traffic engineering shall not be construed to permit the practice of civil, electrical, or mechanical engineering.

404.1. Responsible Charge—Professional Engineering.

(a) As used in the Professional Engineers Act, the term "responsible charge" directly relates to the extent of control a professional engineer is required to maintain while exercising independent control and direction of professional engineering services or creative work and to the engineering decisions which can be made only by a professional engineer.

(1) Extent of Control. The extent of control necessary to be in responsible charge shall be such that the engineer:

(A) Makes or reviews and approves the engineering decisions defined and described in subdivision (a)(2) below.

(B) In making or reviewing and approving the engineering decisions, determines the applicability of design criteria and technical recommendations provided by others before incorporating such criteria or recommendations.

(2) Engineering Decisions. The term "responsible charge" relates to engineering decisions within the purview of the Professional Engineers Act.

Engineering decisions which must be made by and are the responsibility of the engineer in responsible charge are those decisions concerning permanent or temporary projects which could create a hazard to life, health, property, or public welfare, and may include, but are not limited to:

(A) The selection of engineering alternatives to be investigated and the comparison of alternatives for the project.

(B) The selection or development of design standards or methods, and materials to be used.

(C) The decisions related to the preparation of engineering plans, specifications, calculations, reports, and other documents for the engineered works.

(D) The selection or development of techniques or methods of testing to be used in evaluating materials or completed projects, either new or existing.

(E) The review and evaluation of manufacturing, fabrication or construction methods or controls to be used and the evaluation of test results, materials and workmanship insofar as they affect the character and integrity of the completed project.

(F) The development and control of operating and maintenance procedures.

(3) Reviewing and Approving Engineering Decisions. In making or reviewing and approving engineering decisions, the engineer shall be physically present or shall review and approve through the use of communication devices the engineering decisions prior to their implementation.

(b) Responsible Charge Criteria. In order to evaluate whether an engineer is in responsible charge, the following must be considered: The professional engineer who signs engineering documents must be capable of answering questions asked by individuals who are licensed by the Board in the appropriate branch of professional engineering relevant to the project and who are fully competent and proficient by education and experience in the field or fields of professional engineering relevant to the project. These questions would be relevant to the engineering decisions made during the individual's participation in the project, and in sufficient detail to leave little question as to the engineer's technical knowledge of the

engineering performed. It is not necessary to defend decisions as in an adversarial situation, but only to demonstrate that the individual in responsible charge made, or reviewed and approved, them and possessed sufficient knowledge of the project to make, or review and approve, them. Examples of questions to be answered by the engineer could relate to criteria for design, methods of analysis, methods of manufacture and construction, selection of materials and systems, economics of alternate solutions, and environmental considerations. The individual should be able to clearly express the extent of control and how it is exercised and to demonstrate that the engineer is answerable within said extent of control.

(c) Successor Licensee. In situations when the professional engineer in responsible charge of an engineering project is unavailable to complete the project or when the project is a site specific adaptation of a previous design, a professional engineer (hereinafter referred to as the "successor licensee") may assume responsible charge of the project as long as the successor licensee exercises the requisite extent of control and assumes responsibility for the engineering decisions as required by subdivision (a) and meets the criteria described in subdivision (b), as well as meeting the requirements of the Professional Engineers Act and Sections 411 and 415. Except as provided in Sections 6735, 6735.3, and 6735.4 of the Code, the original licensee is not relieved of any responsibility arising from the engineering services of which he or she was in responsible charge.

(d) Portions of Projects. Nothing in this section prohibits a professional engineer from providing services for portions of or to add to or to modify an engineering project engineered under the responsible charge of another licensee as long as the professional engineer exercises the requisite extent of control and assumes responsibility for the engineering decisions as required by subdivision (a) and meets the criteria described in subdivision (b), as well as meeting the requirements of the Professional Engineers Act and Sections 411 and 415. The professional engineer need only be in responsible charge of the portions, additions, or modifications or the portion of the project affected by the addition or modification and not of the entire project. Except as provided in Sections 6735(b), 6735.3(b), and 6735.4(b) of the Code, the original licensee is not relieved of any responsibility arising from the engineering services of which he or she was in responsible charge.

(e) The term "responsible charge" does not refer to any of the following:

(1) the concept of financial liability;

(2) management control in a hierarchy of professional engineers except as each of the individuals in the hierarchy exercises independent engineering judgment and thus responsible charge;

(3) such administrative and management functions as accounting, labor relations, personnel performance standards, marketing of services, or goal setting. While an engineer may also have such duties in this position, it should not enhance or decrease one's status of being in responsible charge of the engineering.

404.2. Responsible Charge—Professional Land Surveying.

(a) The term "responsible charge" directly relates to the extent of control a licensed land surveyor or civil engineer legally authorized to practice land surveying (hereinafter referred to as "legally authorized civil engineer") is required to maintain while exercising independent control and direction of land surveying work or services, and the land surveying decisions which can be made only by a licensed land surveyor or legally authorized civil engineer.

(1) Extent of Control. The extent of control necessary to be in responsible charge shall be such that the land surveyor or legally authorized civil engineer:

(A) Makes or reviews and approves the land surveying decisions defined and described in subdivision (a)(2) below.

(B) In making or reviewing and approving the land surveying decisions, determines the applicability of survey criteria and technical recommendations provided by others before incorporating such criteria or recommendations.

(2) Land Surveying Decisions. The term "responsible charge" relates to land surveying decisions within the purview of the Professional Land Surveyors' Act.

Land surveying decisions which must be made by and are the responsibility of the land surveyor or legally authorized civil engineer in responsible charge are those decisions concerning permanent or temporary work which could create a hazard to life, health, property, or public welfare, and may include, but are not limited to:

(A) Selecting the methods, procedures, and tolerances of field work.

(B) Determining calculation and adjustment methods.

(C) Determining and specifying the information to be shown on maps or documents furnished in connection with land surveying services, including the format of the information and the format of the maps or documents.

(D) The decisions related to the preparation of maps, plats, land surveying reports, descriptions, and other land surveying documents furnished in connection with the land surveying services.

(E) Reviewing the sufficiency and accuracy of the work product.

(3) Reviewing and Approving Land Surveying Decisions. In making or reviewing and approving land surveying decisions, the land surveyor or legally authorized civil engineer shall be physically present or shall review and approve through the use of communication devices the land surveying decisions prior to their implementation.

(b) Responsible Charge Criteria. In order to evaluate whether a person authorized to practice land surveying is in responsible charge, the following must be considered: The land surveyor or legally authorized civil engineer who signs surveying documents must be capable of answering questions asked by licensees of the Board who are fully competent and proficient by education and experience in the field or fields of professional land surveying relevant to the project. These questions would be relevant to the decisions made during the individual's participation in the project, and in sufficient detail to leave little question as to the land surveyor's or legally authorized civil engineer's technical knowledge of the work performed. It is not necessary to defend decisions as in an adversarial situation, but only to demonstrate that the individual in responsible charge made, or reviewed and approved, them and possessed sufficient knowledge of the project to make, or review and approve, them.

Examples of questions to be answered by the land surveyor or legally authorized civil engineer could relate to criteria for measurement, surveying methods, analysis, and conclusions made including, but not limited to, the retracement of government surveys, interpretation and construction of deed descriptions, conflicts between construction drawings and actual conditions, determination of the proper control datum and epoch, application of proportion methods and analysis of evidence related to written and unwritten property rights. The individual shall be able to clearly express the extent of control and how it is exercised and to demonstrate that the land surveyor or legally authorized civil engineer is answerable within said extent of control.

(c) Successor Licensee. In situations when the professional land surveyor or legally authorized civil engineer in responsible charge of a land surveying project is unavailable to complete the project, a professional land surveyor or legally authorized civil engineer (hereinafter referred to as the "successor licensee") may assume responsible charge of the project as long as the successor licensee exercises the extent of control and assumes responsibility for the surveying decisions as required by subdivision (a) and meets the criteria described in subdivision (b), as well as meeting the requirements of the Professional Land Surveyors' Act and Sections 411 and 415. Except as provided in Section 8761.2 of the Code, the original licensee is not relieved of any responsibility arising from the land surveying services of which he or she was in responsible charge.

(d) Portions of Projects. Nothing in this section prohibits a professional land surveyor or legally authorized civil engineer from providing services for portions of or to add to or to modify a land surveying project performed under the responsible charge of another licensee as long as the professional land surveyor or legally authorized civil engineer exercises the requisite extent of control and assumes responsibility for the land surveying decisions as required by subdivision (a) and meets the criteria described in subdivision (b), as well as meeting the requirements of the Professional Land Surveyors' Act and Sections 411 and 415. The professional land surveyor or legally authorized civil engineer need only be in responsible charge of the portions, additions, or modifications or the portion of the project affected by the addition or modification and not of the entire project. Except as provided in Section 8761.2 of the Code, the original licensee is not relieved of any responsibility arising from the land surveying services of which he or she was in responsible charge.

(e) The term "responsible charge" does not refer to any of the following:

(1) the concept of financial liability;

(2) management control in a hierarchy of land surveyors or legally authorized civil engineers except as each of the individuals in the hierarchy exercises independent land surveying judgment and thus responsible charge;

(3) such administrative and management functions as accounting, labor relations, personnel performance standards, marketing of services, or goal setting. While a land surveyor or legally authorized civil engineer may also have such duties in this position, it should not enhance or decrease one's status of being in responsible charge of the work.

405. Delegation of Certain Functions.

(a) Whenever it is stated in these rules that the "Board" may or shall exercise or discharge any power, duty, purpose, function, or jurisdiction, the Board specifically has reserved the same for its own, exclusive action.

(b) Whenever it is stated the "executive officer" may or shall exercise or discharge any power, duty, purpose, function, or jurisdiction, the executive officer of the Board has the authority to act thereon.

(c) Any party in interest may appeal to the Board for review of the actions and decisions of the executive officer.

(d) Nothing herein prohibits the executive officer from redelegating to his/her subordinates as provided in Section 18572 of the Government Code.

(e) The power and discretion conferred by law upon the Board to receive and file accusations; issue notices of hearing, statements to respondent and statements of issues; receive and file notices of defense; determine the time and place of hearings under Section 11508 of the Government Code, issue subpoenas and subpoenas duces tecum, set and calendar cases for hearing and perform other functions necessary to the businesslike dispatch of the business of the Board in connection with proceedings under the provisions of Sections 11400 through 11529 of the Government Code, prior to the hearing of such proceedings; and the certification and delivery or mailing of copies of decisions under Section 11518 of said Code are hereby delegated to and conferred upon the executive officer, or to his/her designee.

407. Fees.

(a) All fees required by provisions of the code as implemented by the board shall be transmitted by money order, bank draft, cash or check, payable to the Department of Consumer Affairs, at Sacramento.

(b) The following is the prescribed application fee for:

(1) Authority to use the title "structural engineer" $275

(2) Authority to use the title "geotechnical engineer" $275

(3) Licensure as a professional engineer $275

(4) Licensure as a professional land surveyor $275

(5) Certification as an engineer-in-training or as a land surveyor-in-training $100

(c) The two-year biennial renewal fee for a license that expires on or after July 1, 2003, shall be $150. The fee for renewal of a license that expires on or after October 1, 2005, shall be $125.

(d) The fee for an examination appeal filed pursuant to Rule 444 shall be $134.00.

(e) The fee for each retired license shall be $87.50; no renewal fee or other fee shall be charged for the retired license. (As used in this subdivision, "license" includes certificate of registration or license as a professional engineer, licensure as a professional land surveyor, and certificates of authority to use the titles "structural engineer," "geotechnical engineer," "soil engineer," "soils engineer," or "consulting engineer.")

(f) Fees required under provisions of this rule transmitted through the United States mail shall be deemed filed on the date shown by the post office cancellation mark stamped on the envelope containing it, or on the date mailed if satisfactory proof is made that mailing occurred on an earlier date.

(g) Renewal applications filed with the Board more than thirty (30) days after 12 midnight on the expiration date pursuant to the Professional Land Surveyors' Act and more than sixty (60) days after 12 midnight on the expiration date pursuant to the Professional Engineers Act and not accompanied by the prescribed delinquent penalty fee equal to 50 percent of the renewal fee, shall be returned by the executive officer with a statement of the reason therefor.

(h) Refund of fees submitted to the Board shall be made only as follows:

(1) Any application fees or penalties imposed and collected illegally, by mistake, inadvertence or error shall be refunded in full.

(2) An applicant for licensure as a professional engineer, for licensure as a professional land surveyor, for permission to use the title "structural engineer" or "geotechnical engineer," for certification as an engineer-in-training or for certification as a land surveyor-in-training found not eligible for admission to the examination requested is entitled to a refund of one-half of the application fee. Upon request, one-half of the application fee shall be refunded to the estate of an applicant who dies prior to taking an examination.

408. Meetings.

(a) The board will meet at times and places within California designated by the board and shall hold at least two regular meetings each year.

(b) Special meetings of the board shall be called from time to time by the Board president when necessary. Special meetings shall also be called by the executive officer upon a written request signed by two board members.

(c) All meetings will be noticed in accordance with the requirements of the Bagley-Keene Open Meeting Act (Government Code Sections 11120 et seq.).

410. Certificates.

(a) Certificates and licenses will be issued in the order in which the applicants qualify.

(b) A duplicate of a certificate issued in accordance with Section 6765 of the Professional Engineers Act or Section 8749 of the Professional Land Surveyors' Act shall be issued only to replace one lost, destroyed, or mutilated, upon a written request accompanied by a fee of $10 and an affidavit verifying the loss, destruction or mutilation of the previous certificate. The affidavit of lost license must be submitted on a form provided by the Board.

411. Seal and Signature.

(a) The seal required by Section 6764 of the Code shall be not less than one and one-half (1½) inches in diameter and shall contain the following information:

(1) Within the top border of seal: Either "Professional Engineer," "Registered Professional Engineer," or "Licensed Professional Engineer."

(2) Within the bottom border of seal: "State of California."

(3) In the center of seal, from top to bottom:

(A) Licensee's name as it appears on the certificate issued by the Board or as abbreviated pursuant to subdivision (d);

(B) Number of certificate or authority;

(C) Expiration date or space within which the expiration date shall be written;

and,

(D) Branch or authority of engineering in which licensed.

The seal shall be of a design similar to those shown below and shall bear at minimum those elements specified above.

(b) The seal authorized by Section 8750 of the Code shall be not less than one and one-half (1 ½) inches in diameter and shall contain the following information:

(1) Within the top border of the seal: Either "Professional Land Surveyor" or "Licensed Land Surveyor."

(2) Within the bottom border of the seal: "State of California."

(3) In the center of the seal, from top to bottom:

(A) Licensee's name as it appears on the certificate issued by the Board or as abbreviated pursuant to subdivision (d);

(B) Number of certificate;

(C) Expiration date or space within which the expiration date shall be written.

The seal shall be of a design similar to those shown below and shall bear at minimum those elements specified above.

(c) The seal may be obtained by the licensee from any source.

(d) The seal may contain an abbreviated form of the licensee's given name or a combination of initials representing the licensee's given name provided the surname listed with the Board appears on the seal and in the signature.

(e) The seal shall be capable of leaving a permanent ink representation, an opaque and permanent impression, or an electronically-generated representation on the documents. The signature may be applied to the documents electronically.

(f) Preprinting of blank forms with the seal or signature, the use of decals of the seal or signature, or the use of a rubber stamp of the signature is prohibited.

(g) (1) All professional engineering plans, specifications, reports, or documents (hereinafter referred to as "documents") shall be signed and sealed in accordance with the requirements of the Professional Engineers Act and any other laws related to the practice of professional engineering and shall be signed and sealed in a manner such that all work can be clearly attributed to the licensee(s) in responsible charge of the work.

(2) All maps, plats, reports, descriptions, or other professional land surveying documents (hereinafter referred to as "documents") shall be signed and sealed in accordance with the requirements of the Professional Land Surveyors' Act and any other laws related to the practice of

professional land surveying and shall be signed and sealed in a manner such that all work can be clearly attributed to the licensee(s) in responsible charge of the work.

(3) When signing and sealing documents containing work done by or under the responsible charge of two or more licensees, the signature and seal of each licensee in responsible charge shall be placed on the documents with a notation describing the work done under each licensee's responsible charge.

(h) Each licensee shall include the date of signing and sealing immediately below or next to the signature and seal.

412. Address Change.

Each person who is an applicant for, or a holder of, a certificate or license issued by the Board under provisions of the Professional Engineers Act or the Professional Land Surveyors' Act shall file his/her address with the Board office. Within thirty (30) days after changing addresses, he/she shall notify the Board office of such change.

415. Practice Within Area of Competence.

A professional engineer or land surveyor licensed under the Code shall practice and perform engineering or land surveying work only in the field or fields in which he/she is by education and/or experience fully competent and proficient.

Nothing in this regulation shall be construed: (1) to prohibit a professional engineer from signing plans which include engineering work in areas other than that in which he/she is fully competent and proficient, if such work was performed by other engineers who were fully competent and proficient in such work; (2) to prohibit a professional engineer from performing engineering work or a land surveyor from performing land surveying work in areas which involve the application of new principles, techniques, ideas or technology; (3) to prohibit a professional engineer from supervising other engineers or a land surveyor from supervising other land surveyors who may respectively be performing engineering work or land surveying work in areas other than those in which

the supervising professional engineer or supervising land surveyor is fully competent and proficient; and (4) to prohibit a professional engineer from signing plans which include engineering work, portions of which were designed or required by any governmental agency.

416. Substantial Relationship Criteria.

For the purpose of denial, suspension, or revocation of the license of a professional engineer or a land surveyor pursuant to Division 1.5 (commencing with Section 475) of the Business and Professions Code, a crime or act shall be considered substantially related to the qualifications, functions, and duties of a professional engineer or land surveyor if, to a substantial degree, it evidences present or potential unfitness of a professional engineer or land surveyor to perform the functions authorized by his or her license in a manner consistent with the public health, safety, or welfare. Such crimes or acts shall include, but not be limited to, those involving the following:

(a) For professional engineers, any violations of the provisions of the Professional Engineers Act or aiding and abetting any person in such a violation;

(b) For land surveyors, any violations of the provisions of the Professional Land Surveyors' Act or aiding and abetting any person in such a violation;

(c) A conviction of a crime arising from or in connection with the practice of professional engineering or land surveying.

418. Criteria for Rehabilitation.

(a) When considering the denial of an application for certification as an engineer-intraining or a land surveyor-in-training, or for licensure as a professional engineer, or for licensure as a professional land surveyor, or for authority to use the title "structural engineer," or for authority to use the title "geotechnical engineer," under Section 480 of the Code, the Board will consider the following criteria in evaluating the rehabilitation of the applicant and his or her present eligibility for such a licensure or authority:

(1) The nature and severity of the act(s) or crime(s) under consideration as grounds for denial.

(2) Evidence of any act(s) committed prior to or subsequent to the act(s) or crime(s) under consideration as grounds for denial which could also be considered as grounds for denial under Section 480 of the Code.

(3) The time that has elapsed since commission of the act(s) or crime(s) referred to in subdivision (1) or (2).

(4) The extent to which the applicant has complied with any terms of parole, probation, restitution, or any other sanctions lawfully imposed against the applicant.

(5) Any evidence of rehabilitation submitted by the applicant.

(6) Total criminal record.

(7) If applicable, evidence of expungement proceedings pursuant to Section 1203.4 of the Penal Code.

(b) When considering the suspension or revocation of the certification of an engineer-intraining or a land surveyor-in-training, or the license of a professional engineer or a professional land surveyor, or the authority to use the title "structural engineer," or the authority to use the title "geotechnical engineer" under Section 490 of the Code, the Board will consider the following criteria in evaluating the rehabilitation of such person and his or her present eligibility to retain his or her license:

(1) The nature and severity of the act(s) or crime(s) under consideration as grounds for suspension or revocation.

(2) Evidence of any act(s) committed prior to or subsequent to the act(s) or crime(s) under consideration as grounds for suspension or revocation which could also be considered as grounds for suspension or revocation under Section 490 of the Code.

(3) The time that has elapsed since commission of the act(s) or crime(s) referred to in subdivision (1) or (2).

(4) The extent to which the licensee has complied with any terms of parole, probation, restitution, or any other sanctions lawfully imposed against the license.

(5) Any evidence of rehabilitation submitted by the licensee.

(6) Total criminal record.

(7) If applicable, evidence of expungement proceedings pursuant to Section 1203.4 of the Penal Code.

(c) When considering a petition for reinstatement of the certification of an engineer-intraining or a land surveyor-in-training, or the license of a professional engineer or a professional land surveyor, or the authority to use the title "structural engineer," or the authority to use the title "geotechnical engineer," the Board shall evaluate evidence of rehabilitation submitted by the petitioner, including but not limited to the following:

(1) Educational courses, including college-level courses, seminars, and continuing professional development courses, completed after the effective date of the Board's decision ordering revocation.

(2) Professional engineering or land surveying work done under the responsible charge of a licensee in good standing or under the direction of a person legally authorized to practice.

(3) Payment of restitution to the consumer(s) by the petitioner.

(4) Actual or potential harm to the public, client(s), employer(s), and/or employee(s) caused by the action(s) that led to the revocation or that could be caused by the reinstatement of the certificate, license, or authority.

(5) The criteria specified in subsection (b)(1) through (7), as applicable.

(6) Disciplinary history, other than criminal actions, after the revocation.

(7) Recognition by the petitioner of his or her own actions and/or behavior that led to the revocation.

(8) Correction of the petitioner's actions and/or behavior that led to the revocation.

419. Disciplinary Orders.

For violations of Business and Professions Code sections 6775 and/or 8780 which result in an order issued in accordance with Chapters 4.5 and 5 of Part 1 of Division 3 of Title 2 of the Government Code against a professional engineering and/or a professional land surveying license, the following provisions shall apply to disciplinary orders contained in decisions of the Board:

(a) The minimum disciplinary order shall be reproval. The maximum disciplinary order shall be revocation of the license.

(b) If warranted by extenuating and/or mitigating factors in the matter, the disciplinary order may be stayed by an express condition that the respondent comply with probationary conditions. The minimum time period in which the respondent shall have to comply with the conditions shall be two years. For purposes of this section, this time period shall be known as the "period of probation."

(c) All decisions containing stayed disciplinary orders as described in subdivision (b) shall include the following probationary conditions:

(1) The respondent shall obey all laws and regulations related to the practices of professional engineering and professional land surveying.

(2) The respondent shall submit such special reports as the Board may require.

(3) The period of probation shall be tolled during the time the respondent is practicing exclusively outside the state of California. If, during the period of probation, the respondent practices exclusively outside the state of California, the respondent shall immediately notify the Board in writing.

(4) If the respondent violates the probationary conditions in any respect, the Board, after giving the respondent notice and the opportunity to be heard, may vacate the stay and reinstate the disciplinary order which was stayed. If, during the period of probation, an accusation or petition to vacate stay is filed against the respondent, or if the matter has been submitted to the Office of the Attorney General for the filing of such, the Board shall have continuing jurisdiction until all matters are final, and the period of probation shall be extended until all matters are final.

(5) Upon successful completion of all of the probationary conditions and the expiration of the period of probation, the respondent's license shall be unconditionally restored.

(d) All decisions containing stayed disciplinary orders as described in subdivision (b) may include one or more of the following probationary conditions:

(1) The respondent's license shall be suspended for a period not to exceed two years. If a suspension of the license is ordered, it shall begin on the effective date of the decision.

(2) Within 60 days of the effective date of the decision, the respondent shall successfully complete and pass the California Laws and Board Rules examination, as administered by the Board.

(3) The respondent shall successfully complete and pass a course in professional ethics, approved in advance by the Board or its designee. The probationary condition shall include a time period in which this course shall be successfully completed which time period shall be at least 60 days less than the time period ordered for the period of probation.

(4) Within 30 days of the effective date of the decision, the respondent shall provide the Board with evidence that he or she has provided all persons or entities with whom he or she has a contractual or employment relationship such that the relationship is in the area of practice of professional engineering and/or professional land surveying in which the violation occurred with a copy of the decision and order of the Board and shall provide the Board with the name and business address of each person or entity required to be so notified. During the period of probation, the respondent may be required to provide the same notification to each new person or entity with whom he or she has a contractual or employment relationship such that the relationship is in the area of practice of professional engineering and/or land surveying in which the violation occurred and shall report to the Board the name and address of each person or entity so notified.

(5) The respondent shall provide verifiable proof to the Board that restitution has been paid as ordered. The probationary condition shall include a time period in which the verifiable proof shall be provided to the Board which time period shall be at least 60 days less than the time period ordered for the period of probation.

(e) In addition to the conditions as may be ordered pursuant to subdivisions (c) and/or (d), the following conditions shall be included for the following specific violations:

(1) Incompetency in the practice of professional engineering and/or professional land surveying:

(A) The respondent shall successfully complete and pass, with a grade of "C" or better, a minimum of one and a maximum of three college-level courses, approved in advance by the Board or its designee. Such courses shall be specifically related to the area of violation. For purposes of this subdivision, "college-level course" shall mean a course offered by a community college or a four-year university of three semester units or the equivalent; "college-level course" does not include seminars. The probationary condition shall include a time period in which the course(s) shall be successfully completed which time period shall be at least 60 days less than the time period ordered for the period of probation.

(B) The respondent shall take and achieve the passing score as set by the Board for the second division examination (including the seismic principles and engineering surveying examinations for civil engineers), provided that in the event the respondent holds multiple licenses, the Board shall select the examination in the area of practice of professional engineering and/or professional land surveying in which the violation occurred and in the area of professional engineering and/or professional land surveying in which the respondent is licensed. The Board or its designee may select the specific examination questions such that the questions relate to the specific area of violation and comprise an examination of the same duration as that required of an applicant for licensure. The respondent shall be required to pay the application fee as described in Section 407 and shall be afforded all examination appeal rights as described in Sections 407, 443, and 444. The probationary condition shall include a time period in which the examination(s) shall be successfully completed which time period

shall be at least 60 days less than the time period ordered for the period of probation.

(C) During the period of probation, the respondent may practice professional engineering and/or professional land surveying only under the supervision of a professional engineer and/or professional land surveyor licensed in the same branch as the respondent. This person or persons shall be approved in advance by the Board or its designee. Such supervising professional engineer and/or professional land surveyor shall initial every stamped or sealed document in close proximity to the respondent's stamp or seal.

(2) Negligence in the practice of professional engineering and/or professional land surveying:

(A) The respondent shall successfully complete and pass, with a grade of "C" or better, a minimum of one and a maximum of three college-level courses, approved in advance by the Board or its designee. Such courses shall be specifically related to the area of violation. For purposes of this subdivision, "college-level course" shall mean a course offered by a community college or a four-year university of three semester units or the equivalent; "college-level course" does not include seminars. The probationary condition shall include a time period in which the course(s) shall be successfully completed which time period shall be at least 60 days less than the time period ordered for the period of probation.

(3) Violation and/or breach of contract in the practice of professional engineering and/or professional land surveying:

(A) The respondent shall successfully complete and pass, with a grade of "C" or better, a minimum of one and a maximum of three college-level courses, approved in advance by the Board or its designee. Such courses shall be specifically related to the area of violation. For purposes of this subdivision, "college-level course" shall mean a course offered by a community college or a four-year university of three semester units or the equivalent; "college-level course" does not include seminars. The probationary condition shall include a time period in which the course(s) shall be successfully completed which time period shall be at least 60 days less than the time period ordered for the period of probation.

(4) Failure to file a record of survey and/or corner record in the practice of professional land surveying:

(A) For any records of survey and/or corner records found not to have been filed and recorded, the respondent shall file or record, as appropriate, the required record(s) with the appropriate governmental agency within 90 days of the effective date of the decision. The respondent shall provide the Board with verifiable proof that the required record(s) have been filed or recorded, as appropriate, by the governmental agency within 30 days of such filing or recordation. If an actual suspension of the respondent's license is ordered as a probationary condition, the record(s) required by this subdivision shall be the only professional land surveying work the respondent is allowed to perform during the suspension.

(f) If the respondent is a civil engineer who is legally authorized to practice professional land surveying and the violation involves negligence and/or incompetency in the practice of professional land surveying and if warranted by aggravating factors in the matter, the disciplinary order shall include the following condition:

(1) The existing civil engineer license shall be revoked; a new civil engineer license shall be issued which does not authorize the respondent to practice professional land surveying. Before being permitted to practice professional land surveying, the respondent shall complete and comply with all of the legal requirements for licensure as a professional land surveyor, including, but not limited to, supplying the appropriate application documents and fees and taking and passing the entire second-division examination in professional land surveying. In addition to the disciplinary orders described in this section, all decisions shall address recovery of the Board's investigation and enforcement costs, as described in and authorized by Business and Professions Code section 125.3.

Notwithstanding this section, non-conforming terms and conditions may be included as part of the disciplinary order, including such other further or lesser action as the Board deems appropriate, in the interest of protecting the public health, safety, and welfare.

As used in this section, "license" includes certificate of registration or license as a professional engineer, licensure as a professional land surveyor, and certificates of authority to use the titles "structural engineer," "geotechnical engineer," "soil engineer," "soils engineer," or "consulting engineer."

ARTICLE 2. APPLICATIONS

420. Applications.

(a) Applications for certification, for licensure, or for a certificate of authority shall be:

(1) Filed on a form prescribed by the executive officer and shall be typewritten.

(2) Filed at the office of the Board and accompanied by the required application fee.

(3) Made out properly in every respect and must contain full information.

(4) Subscribed and certified to "under penalty of perjury" as provided by Section 2015.5 of the Code of Civil Procedure.

(b) An application made otherwise will not be accepted by the Board and it may be returned by the executive officer with a statement of the reason therefor.

(c) Upon evaluation of the applicant's qualifications, his/her examination results and any other supporting data, his/her application will be either:

(1) Denied without prejudice, and the application fee retained by the Board for the Professional Engineers' Fund.

(2) Approved, and he/she will be granted the certification for which application was made.

(d) The Board may request each applicant to provide the Board with a current photograph after an applicant has become licensed.

421. Refile Application.

(a) The executive officer may prescribe a short application form for use of those applicants who, failing an examination, apply within a reasonable period of time after the date of the examination previously failed, for re-examination. This application form may be known as a refile application form. The applicant and his application for re-examination shall be subject to the same provisions of the code and rules of the board, whenever applicable, as govern the filing of an original application.

(b) The applicant for re-examination shall be assigned by the executive officer to the next scheduled examination for which his/her application qualifies him/her.

422. Final Filing Date.

(a) An application filed with the board after the final filing date announced for an examination shall not be considered for such examination.

(b) Whenever the final filing date announced for an examination falls upon a Saturday, Sunday or holiday, it shall be extended to the next business day following.

(c) If an application is filed with the board through the United States mail, it shall be deemed filed on the date shown by the post office cancellation mark stamped on the envelope containing it, or on the date mailed if satisfactory proof is made that the mailing occurred on an earlier date.

424. Experience Requirements—Professional Engineers.

(a) The branches and title authorities described in Section 404, herein, overlap and some activities are common to two or more branches and title authorities. The minimum number of years of qualifying experience in such overlapping branches and title authorities may be used in securing licensure in any applicable branch or title authority but cannot be used more than once. The only exception to this is experience credit for education and experience credit used to qualify for the land surveyor examination. Qualifying education entitles a candidate to experience credit and this experience credit can be used again even though it has already been used to qualify for another examination.

(b) An applicant for licensure as a professional engineer shall be granted four years experience credit for graduation from an approved engineering curriculum; or, two years experience credit for graduation from a non-approved engineering curriculum or an approved engineering technology curriculum. The additional actual work experience required to meet the six years experience requirement shall have been gained after graduation, except for cooperative work-study experience. A maximum of five years experience credit shall be granted for graduation from a cooperative work-study engineering curriculum accredited by the Accreditation Board for Engineering and Technology (ABET).

A graduate of an approved postgraduate engineering curriculum shall be given one year of additional experience credit over and above credit given for undergraduate education. Incomplete undergraduate engineering education at an approved institution shall be given one-half year of experience for each year of study completed.

The sum of qualifying experience credit for education and engineering teaching experience shall not exceed five years.

(c) Qualifying experience is that experience satisfactory to the Board which has been gained while performing engineering tasks under the direction of a person legally qualified to practice in an applicants' branch of engineering.

(1) For the purposes of this section, "legally qualified" means having an appropriate license as a professional engineer, or by being an employee of the Federal Government, or, except for civil engineers, by virtue of being an employee of a manufacturing, mining, public utility, research and development, or other industrial corporation; or by holding an appropriate license as a contractor.

(2) Qualifying experience shall be computed on an actual time worked basis, but not to exceed forty hours per week.

(3) Applied engineering research is an engineering task for the purposes of determining qualifying experience.

(d) Computation of qualifying experience for licensure as a professional engineer or for authority to use the title "structural engineer" or "geotechnical engineer" shall be to the date of filing of the application; or it shall be to the final filing date announced for the examination if the application is filed within a period of thirty (30) days preceding the final filing date announced for such examination.

424.5. Reinstatement Requirements for Delinquent Applicants.

(a) A license which has not been renewed within the time required under Business and Professions Code section 6796.3 or 8803 is considered delinquent and, except as provided in subdivision (c), shall be reinstated if the applicant complies with the following:

(1) Submits evidence satisfactory to the Board that the applicant is qualified in the branch for which he or she is applying. This evidence shall consist of:

(A) A completed, typewritten application on a form as specified in Section 420 accompanied by the required application fee as specified in Section 407 (b)(1)-(4).

(B) Completed appropriate reference forms as specified in Sections 427.10, 427.20, or 427.30. The submission of a reference which states that the applicant is not technically qualified to be licensed shall be grounds for denial.

(2) Takes and passes the examination on the applicable state laws and board regulations as specified in Business and Professions Code section 6755.2 or 8741.1 .

(3) Takes and passes examinations on seismic principles and engineering surveying, if he or she is a civil engineering applicant whose initial registration was issued prior to January 1, 1988.

(4) Pays all accrued and unpaid renewal fees.

(5) Has not committed any acts or crimes constituting grounds for denial of licensure under Business and Professions Code section 480.

(b) An applicant who is unable to submit evidence satisfactory to the Board that he or she is qualified as provided in subdivision (a)(1) shall take and pass the appropriate second division examination or the appropriate title authority examination in addition to

the requirements specified in subdivision (a)(2)- (5) prior to reinstatement of the delinquent license.

(c) Notwithstanding subdivisions (a) and (b), the Board may pursue action, including but not limited to revocation or suspension of the license pursuant to Business and Professions Code sections 6775, 6776, 8780, and 8781, issuance of a citation containing an order to pay an administrative fine pursuant to Sections 473 through 473.4, filing of criminal charges pursuant to Business and Professions Code sections 6787 and 8792, and denial of the application pursuant to Section 420, against the applicant if evidence obtained during an investigation reveals that the applicant has violated any provision of the Business and Professions Code, the California Code of Regulations, or other applicable laws and regulations related to the practices of professional engineering or professional land surveying during the period of delinquency, including, but not limited to, practicing or offering to practice with an expired or delinquent license.

(d) The application response timeframe is as specified in Section 470 (a).

(e) The Board's time period for processing an application from receipt of the initial application to the final decision regarding issuance or denial of licensure is as specified in Section 471.

As used in this section, "license" includes certificate of registration as a professional engineer, licensure as a professional land surveyor, and certificates of authority to use the title "structural engineer," "soil engineer," or "consulting engineer."

425. Experience Requirements—Professional Land Surveyors.

(a) An applicant for licensure as a professional land surveyor shall fulfill the educational and experience requirements contained in Sections 8741 and 8742 of the Code.

(b) All qualifying work experience in land surveying shall be performed under the direction and review of a person legally authorized to practice land surveying. An applicant shall possess at least two years of actual responsible training experience in land surveying which shall involve at least four of the land surveying activities specified in subdivisions (a)-(g) and (k)-(m) of Section 8726 of the Code. Qualifying experience in activities specified in subdivision (a), (b), and (m) of Section 8726 shall not exceed one year. Qualifying experience shall be computed on an actual time worked basis, but not to exceed forty hours per week.

(c) An applicant shall be credited with qualifying experience for post-secondary education that may be applied to the six years necessary for admission to the professional land surveyor examination. A graduate from a four-year curriculum with an emphasis in land surveying accredited by the Accreditation Board for Engineering and Technology (ABET) shall be given four years experience credit. A graduate of a curriculum with an emphasis in land surveying not accredited by ABET shall be given two years experience credit. Incomplete undergraduate education in land surveying at an institution accredited by ABET shall be given one-half year of experience for each year of study completed, except that the maximum of such experience credit shall be two years per applicant. A year of study shall be at least 32 semester units or 48 quarter units, no less than 10 semester units or 15 quarter units of which shall be from classes clearly identified as being land surveying subjects. A maximum of five years experience shall be credited for graduation from a cooperative work-study land surveying curriculum accredited by the Accreditation Board for Engineering and Technology (ABET).

(d) For purposes of Section 8742 of the Code, the term "responsible field training" experience may include, but is not limited to, the land surveying activities listed below. Under the responsible charge, direction, and review of a person legally authorized to practice land surveying, the applicant:

(1) Determines field survey methods and procedures, including selection of accuracy standards.

(2) Selects or verifies that the correct control monumentation is used to establish the designated survey datum(s) (horizontal and vertical) and selects on-the-ground locations for control monuments.

(3) Determines the relevance of monuments and physical field evidence for the purpose of establishing boundary and property lines.

(4) Reviews measurement observations for the determination of accuracy, completeness, and consistency.

(5) Reviews field notes and records for application of proper field survey procedures.

(6) Plans, performs, and reviews field checks and, based on such checks, determines if completed field surveys are accurate and sufficient.

(7) Searches for boundary and control monuments; assists in analyzing field evidence for locating boundary points and lines; identifies and describes such evidence; compares record data to found physical evidence; compares record data to measured data; documents discrepancies; assists in acquiring and documenting testimony regarding boundary locations; recommends boundary location and/or establishment; selects or verifies that the correct controlling monuments are used to locate or establish boundary points and lines; and prepares draft record documents.

(8) Coordinates the fieldwork necessary to prepare maps, plats, reports, descriptions, or other documents.

(9) Recommends when existing boundary monuments are to be replaced, selects the method(s) to be used for replacing and resetting monuments, and prepares field documentation of such work, including that necessary for Parcel Maps, Final Maps, Record of Survey Maps, and Corner Records.

(10) Functions as a party chief, chief of parties, or lead person in charge of field crew(s) in the performance of field surveys.

(11) Plans and performs field observations using Global Positioning System technology and determines if completed field surveys are accurate and sufficient in geodetic and land surveying applications.

(12) Performs surveys to facilitate the location or construction of infrastructure and fixed works of improvement. The enumeration of the above tasks does not preclude the Board from awarding "responsible field training" credit for training of a similar character in other current or future land surveying activities not specifically enumerated herein. It is also understood that the listed tasks are only some of those that may be considered as responsible training, and that this list is not in any way intended to enumerate all of the tasks which may be performed by licensed Professional Land Surveyors.

(e) For purposes of Section 8742 of the Code, the term "responsible office training" experience may include, but is not limited to, the land surveying activities listed below. Under the responsible charge, direction, and review of a person authorized to practice land surveying, the applicant:

(1) Performs the planning and analysis necessary for the preparation of survey documents, such as Parcel Maps, Final Maps, Record of Survey Maps, Corner Records, legal descriptions, topographic maps, plat maps, lot line adjustments, annexations, and boundary line agreements.

(2) Reduces and evaluates field data.

(3) Develops procedures and systems for the collection, reduction, adjustment, and use of land surveying data.

(4) Prepares data to be used by field surveyors or field crews.

(5) Coordinates the processing of maps, plats, reports, descriptions, or other documents with local agencies, other licensed surveyors, or County Surveyors Offices.

(6) Coordinates the office work necessary to prepare maps, plats, reports, descriptions, or other documents.

(7) Coordinates survey and design efforts for improvement plans as required for sufficiency to enable proper location of improvements in the field.

(8) Researches public and private records to obtain survey and title data.

(9) Performs boundary analysis and determination using record descriptions, survey, and title data.

(10) Plans and coordinates the application of Global Positioning System technology for geodetic and land surveying applications.

(11) Plans, coordinates, performs, and reviews the entry of property boundary related geo-referenced data into an electronic database.

(12) Prepares topographic mapping utilizing photogrammetric methods.

The enumeration of the above tasks does not preclude the Board from awarding "responsible office training" credit for training of a similar character in other current or future land surveying activities not specifically enumerated herein. It is also understood that the listed tasks are only some of those that may be considered as responsible training, and that this list is not in any way intended to enumerate all of the tasks which may be performed by licensed professional land surveyors.

(f) Computation of qualifying experience for a license as a professional land surveyor shall be to the date of filing of the application, or it shall be to the final filing date announced for the examination if the application is filed within a period of thirty (30) days preceding the final filing date announced for such examination.

(g) An applicant for licensure as a land surveyor who holds a valid and unexpired registration or license as a civil engineer is exempt from the application requirements of this section provided he or she submits sufficient documentation that he or she has a minimum of two years of actual experience in land surveying as required by Business and Professions Code Section 8742(a)(3).

426.10. Qualification Requirements for Structural Authority.

An applicant for authority to use the title "structural engineer" shall comply with all of the following requirements:

(a) The applicant shall hold an unexpired, valid California license as a civil engineer.

(b) The applicant shall submit evidence satisfactory to the Board that the applicant has been in responsible charge of structural engineering qualifying experience, as defined in Section 426.11 and/or Section 426.12, for a minimum of three years subsequent to the date of examination which was passed to gain California license as a civil engineer or as provided in Section 426.14.

426.11. Qualifying Experience for Structural Authority.

"Structural Engineering qualifying experience" is defined as acceptable professional practice in responsible charge of structural engineering projects as related to buildings (or other structures) and shall include structural design experience in all areas as specified in subdivisions (a)-(f) below because the stability of a structure is dependent upon the interaction of the individual structural components as well as the structure as a whole:

(a) Common Construction Materials - Steel, Concrete, Wood and Masonry: A structural engineer shall have experience in the use of three of the four common construction materials of steel, concrete, wood, and masonry as they relate to the design, rehabilitation and/or investigation of buildings (or other structures);

(b) Determination of Lateral Forces: A structural engineer shall have experience regarding structural design to resist lateral forces;

(c) Selection of Framing Systems: A structural engineer shall have experience regarding the selection of framing systems, including the consideration of alternatives and the selection of an appropriate system for the interaction of structural components to support vertical and lateral loads;

(d) Selection of Foundation Systems: A structural engineer shall have experience in the selection of foundation systems, including the consideration of alternatives and the selection of an appropriate type of foundation system to support the structure;

(e) Application of Code Requirements: A structural engineer shall have experience in applying local, state and federal requirements relating to design loads, materials, and detailing; and

(f) Multi-story Buildings or Equivalent Multi-level Structures: A structural engineer shall have experience with the design and detailing for the transfer of forces between stories in multi-story buildings. A multi-story building is a building which is more than one story in height and which is not exempted pursuant to Section 6737.1 of the code.

426.12. Experience for Checking Structural Plans.

The Board shall consider the following experience as structural engineering qualifying experience, in lieu of that experience defined in Section 426.11: Professional level employment performing the checking of structural engineering plans and calculations, when performed under the immediate supervision of, and certified to by, either a civil engineer who holds a valid California license with the authority to use the title "structural engineer" in this state or a Professional Engineer who is authorized to use the title "structural engineer" registered or licensed outside of this state but registered or licensed in a state which has a comity agreement with the State of California related to structural engineering. However, an applicant who applies for authority to use the title "structural engineer" under this section shall furnish the Board with a verification of employment from each employer which lists the name(s) of the immediate supervisor of the applicant during the period of employment used as qualifying experience under this section.

426.13. Supplemental Evidence of Responsible Charge for Structural Authority.

(a) The board shall consider the following as supplemental evidence, if submitted for consideration, to assist in determining whether an applicant for structural authority possesses the requisite three (3) years of structural engineering qualifying experience at the level of responsible charge as required in Section 426.10 and defined in Sections 426.11 and 426.12:

(1) Project management experience:

(A) Coordination with other disciplines such as civil, electrical and mechanical engineers and/or architects;

(B) Production of construction document packages such as calculations, drawings and specifications; and,

(C) Supervision and/or coordination of staff.

(2) Field experience:

(A) Familiarity with techniques, methods and means of construction;

(B) Field observation of construction for compliance to drawings and specifications; and,

(C) Field investigation of existing structures for evaluation or forensic purposes.

(b) Notwithstanding subsections (a)(1) and (a)(2), other types of experience deemed equivalent to project management or field experience may be considered on a case-by-case basis to assist in determining whether an applicant possesses the three (3) years of structural engineering qualifying experience at the level of responsible charge as required in Section 426.10.

(c) Any experience submitted pursuant to this section shall not be considered as a substitute for the mandatory types of qualifying experience required by Section(s) 426.11 and/or 426.12.

426.14. Experience for Structural Engineering Gained Out of State.

(a) The Board may consider an application for authority to use the title "structural engineer" from an applicant who does not possess three (3) years of qualifying experience subsequent to the date of the examination which was passed to gain licensure as a California civil engineer but who possesses experience equivalent to that provided in Section 426.11 based upon either:

(1) A minimum of three (3) years of structural engineering qualifying experience gained after the applicant's registration or licensure as civil engineer in another state.

(2) A minimum of three (3) years of structural engineering qualifying experience which was gained while exempt from licensure pursuant to Section 6739 of the Code or while employed or registered or licensed in another country. Such experience shall be in addition to the experience required for licensure as a civil engineer in this state.

(b) Applicants seeking approval of their structural engineering qualifying experience, pursuant to this section, shall file their application at least six months prior to the final filing deadline to be considered for the next scheduled examination. Applicants may be required to appear for an interview regarding their structural engineering qualifying experience.

426.50. Qualification Requirements "Soil Engineer."

An applicant for authority to use the title "soil engineer" shall:

(a) Hold an unexpired, valid California license as a civil engineer.

(b) Submit evidence satisfactory to the Board that the minimum number of years of qualifying experience or education has been met as required in Sections 6736.1(b) and 6763 of the Code and as defined in Section 426.51, subsequent to the date of examination which was passed to gain licensure as a civil engineer. In addition, up to one year credit as qualifying experience in responsible charge will be given for possession of post graduate degree(s) from a Board approved school of engineering with major studies in soil engineering as listed in Section 426.51(c). Credit for post graduate degree(s) will not be given if it has already been applied to the experience requirement for civil engineering licensure.

426.51. Qualifying Experience for "Soil Engineer."

"Qualifying experience" means responsible charge of soil engineering projects. Evidence shall be provided that the applicant has qualifying experience in the areas described in subdivisions (a), (c) and (e) and has demonstrated working knowledge in the areas described in subdivisions (a) through (e). At least one-half of the applicant's annual full-time professional practice shall be in soil engineering, except that a teacher of soil engineering and related courses at a board approved school of engineering will be given credit for applicable consulting work as a percentage of equivalent full-time work. Applicable consulting work shall be substantiated by references and project documents.

(a) Development of programs of geotechnical investigation which includes, but is not limited to:

(1) Communication with other design consultants to determine their geotechnical input needs;

(2) Performance of literature searches, site history analyses, etc., related to surface and subsurface conditions;

(3) Formulation or engineering evaluation of field exploration and laboratory testing programs to accomplish the scope of the investigation;

(4) Preparation or engineering evaluation of proposals.

(b) Performance of geotechnical field and laboratory studies which includes, but is not limited to:

(1) Direction and/or modification of field exploration programs, as required upon evaluation of the conditions being encountered;

(2) Classification and evaluation of subsurface conditions.

(3) Understanding the purposes for and being qualified to perform routine field and laboratory tests for:

(A) soil strength

(B) bearing capacity

(C) expansion properties

(D) consolidation characteristics

(E) soil collapse potential

(F) erosion potential

(G) compaction characteristics

(H) material acceptability for use in fill

(I) pavement support qualities

(J) freeze-haw properties

(K) grain-size

(L) permeability/percolation properties

(c) Analysis of geotechnical data and engineering computations which includes, but is not limited to:

(1) Analysis of field and laboratory test results regarding:

(A) soil strength

(B) bearing capacity

(C) expansion properties

(D) consolidation characteristics

(E) soil collapse potential

(F) erosion potential

(G) compaction characteristics

(H) material acceptability for use in fill

(I) pavement support qualities

(J) freeze-thaw properties

(K) grain-size

(L) permeability/percolation properties

(M) ground water conditions

(N) soil dynamic properties

(2) Performance of computations using test results and available data regarding:

(A) bearing capacity

(B) foundation type, depth, dimensions

(C) allowable soil bearing pressures

(D) potential settlement

(E) slope stability

(F) retaining systems

(G) soil treatment

(H) dewatering/drainage

(I) floor support

(J) pavement design

(K) site preparation

(L) fill construction

(M) liquefaction potential

(N) ground response to seismic forces

(O) ground water problems; seepage

(P) underpinning

(d) Performance or engineering evaluation of construction, postconstruction and site monitoring which includes, but is not limited to:

(1) Performance or supervision of geotechnical testing and observation of site grading;

(2) Analysis, design and evaluation of instrumentation programs to evaluate or monitor various phenomena in the field, such as settlement, slope creep, porewater pressures and ground water variations;

(3) Geotechnical observation during construction and/or installation, including but not limited to, spread foundations, drilled piers, piles, slurry walls, anchors, bulkheads, shoring, underpinning and subdrains;

(4) Engineering evaluation of soil related distress.

(e) Preparation or engineering evaluation of geotechnical reports which includes, but is not limited to:

(1) Preparation of appropriate plans, logs, test results and other exhibits;

(2) Documentation of testing and observation;

(3) Preparation of written reports which present findings, conclusions and recommendations of the investigation;

(4) Preparation of specifications and guidelines for achieving the intent of subdivision (e) (3), above.

427.10. References for Professional Engineers and Land Surveyors.

To assist the Board in evaluating qualifications, each applicant shall submit completed reference forms, using form 01A-1B (Rev. 2/86) for professional engineers and using form 01A-19 (Rev. 5/86) for professional land surveyors, from as many references as may be consistent with the length and character of the professional experience, provided the applicant shall not furnish less than the number of references required hereafter:

(a) An applicant for a license as a professional land surveyor or as a professional engineer shall refer to not less than four persons who are authorized to practice in the discipline for which the applicant is applying and who have personal knowledge of the applicant's qualifying experience, none of whom is a relative either by birth or marriage.

(b) Nothing herein contained shall be construed to limit authority of the Board to seek such other information pertinent to the education and experience of the applicant as may be required to verify his or her qualifications. The Board may waive the requirement that only registered or licensed individuals give references for the applicants in disciplines other than civil engineering or land surveying when the applicants have no association with registered or licensed individuals in their work environment.

427.20. Reference Requirements for "Soil Engineer."

(a) An applicant for authority to use the title "soil engineer" shall submit at least four completed reference forms from individuals who hold or held current, valid, unexpired California licenses as civil engineers during the time of the applicant's experience. None of the references shall be related to the applicant by birth or marriage. At least two of these individuals shall be civil engineers who are or were actively engaged in the practice of "soil engineering." Each civil engineer providing a reference shall clearly indicate areas of personal knowledge of the applicant's qualifying experience. Reference forms completed by civil engineers registered or licensed outside of California, in lieu of or in addition to California references, will be considered; however, the Board may require additional information as specified in Section 427.20(d). Reference forms completed by civil engineers registered or licensed outside of the State of California shall be notarized. Information submitted by references is confidential.

(b) Notwithstanding Section 427(a), a reference form shall be submitted for each period of qualifying experience listed on the engagement record form for which the applicant desires credit.

(c) An applicant will be required to verify employment inclusive dates for each period of qualifying experience. Employment verification forms may be used for this purpose.

(d) Nothing contained in this section shall limit the authority of the Board to require that an applicant submit additional references, employment

verifications and other information pertinent to education or experience to verify that the applicant has met the minimum qualifications as defined in Sections 6736.1(a) and (c) of the Code and Sections 426.50 and 426.51.

427.30. References for Structural Authority.

(a) An applicant for authority to use the title "structural engineer" shall submit at least three completed reference forms, using form 01A-9 (Rev. 6/86), from individuals who hold current, valid California licenses as civil engineers, and who are authorized by the Board to use the title "structural engineer," or equivalent thereto, none of whom is related to the applicant by birth or marriage. Each reference shall have personal knowledge of the applicant's qualifying experience and shall have examined the applicant's work. It is preferred that at least one of the references has been a direct supervisor for a period of not less than six months.

(b) "Equivalent thereto" as used in this section, means a professional engineer who is authorized to use the title "structural engineer" in a state which has a comity agreement with this state related to "structural engineering." Reference forms completed by a "structural engineer" registered outside of this state but registered or licensed in a state which has a comity agreement with the State of California shall be notarized.

(c) Nothing contained in this section shall limit the authority of the Board to require that an applicant submit additional references, employment verifications and other information pertinent to the applicant's education and/or experience to verify that the applicant meets the minimum qualifications as defined in Sections 426.10, 426.11 and/or 426.13.

428. Abandoned Applications.

In the absence of special circumstances any of following actions by an applicant for certification or licensure shall be considered to constitute abandonment of the application and shall result in cancellation of the application with no refund of the filing fee:

(a) Failure to provide additional information or references within 90 days following the mailing of a request by the Board's staff; or

(b) Failure to complete that examination to which the application has been assigned within two (2) years from the date of filing of the application; or

(c) Failure to appear for examination at the designated time and place unless a postponement has been obtained in accordance with Rule 446; or

(d) Failure to appear for examination at the designated time and place after having obtained two postponements.

429. Application Appeal.

(a) An applicant who is notified by the board that his/her application has been denied may appeal to the board for re-evaluation of his/her application. An application appeal shall be filed with the board within 60 days after the date the denial notice has been mailed to him/her.

(b) An application appeal shall be made in writing and shall state the reason therefor. An appeal shall be supported by additional evidence, more references, affidavits, and supplemental information such that the board may be better informed of the applicant's qualifications.

(c) The executive officer may deny an application appeal which is not filed within the time period provided in paragraph (a) of this rule.

(d) The executive officer shall notify each applicant who appeals under this rule of the approval of his/her appeal, or the reason for its denial.

(e) When an application has been denied, the executive officer shall also notify the applicant that he or she has the right to hearing under the Administrative Procedure Act (Government Code Section 11500 et seq.), if he or she makes a written request for hearing within 60 days after service of the notice of denial.

ARTICLE 3. EXAMINATIONS

436. Schedule of Regular Written Examinations.

(a) Written examinations shall be given at intervals as determined by the board but not less than once each year.

(b) The executive officer shall publish annually, not later than three months prior to the end of each calendar year, a schedule of examinations for the following year.

(c) Whenever circumstances warrant such action the board may postpone, advance, or otherwise change without notice the examination schedule previously published.

437. Individual Examination.

(a) Individual examinations for licensure, certification, or authorization shall be either oral or written or a combination of both, in the discretion of the Board. They may be held at times and places convenient to the Board.

(b) An applicant for licensure as a professional engineer or as a land surveyor will be considered for assignment to an individual examination provided his/her qualifications meet all the requirements of the code and rules of the Board and provided he/she holds a valid registration or license as a professional engineer as a land surveyor in another state, in the same branch in which he/she is applying; such registration or license having been obtained by passing written examinations of comparable standard to those examinations required in California.

(c) An applicant for authority to use the title "structural engineer" may be considered for the individual examination only if he/she has passed a 16-hour examination in another state which the Board may deem equivalent to the written examination for structural authority given in California. This 16-hour examination shall have been in addition to the regular examination series for registration or licensure as a professional engineer and shall have contained significant emphasis on seismic design and lateral load considerations.

438. Waiver of Fundamentals Examination.

(a) An applicant for licensure as a professional engineer whose qualifications meet all requirements of the code and rules of the Board will be allowed to appear for only the second division of the written examination prescribed by Section 6755 of the

Code if he or she meets one or more of the following requirements:

(1) Holds valid licensure as a professional engineer in another branch in California.

(2) Holds valid certification as an engineer-in-training in another state obtained by passing a written examination which normally requires a minimum of eight hours to complete and the content of the examination is designed to test the candidates knowledge of fundamental engineering subjects, including mathematics and the basic sciences.

(3) Is a graduate of an approved engineering curriculum and submits satisfactory evidence to the Board that he or she has fifteen (15) years or more of additional engineering work experience satisfactory to the Board that has been gained in addition to graduation, or any other education experience, and while performing engineering tasks under the direction of a person legally authorized to practice in an applicant's branch of engineering.

(4) Is a graduate of a nonapproved engineering curriculum or an approved engineering technology curriculum and submits satisfactory evidence to the Board that he or she has seventeen (17) years or more of additional engineering work experience satisfactory to the Board that has been gained in addition to graduation, or any other education experience, and while performing engineering tasks under the direction of a person legally authorized to practice in an applicant's branch of engineering.

(5) Is a graduate of an approved engineering curriculum and an approved postgraduate engineering curriculum and submits satisfactory evidence to the Board that he or she has fourteen (14) years or more of additional engineering work experience satisfactory to the Board that has been gained in addition to graduation, or any other education experience, and while performing engineering tasks under the direction of a person legally authorized to practice in an applicant's branch of engineering.

(6) Is a graduate of a nonapproved engineering curriculum or an approved engineering technology curriculum and an approved postgraduate engineering curriculum and submits satisfactory evidence to the Board that he or she has sixteen (16) years or more of additional engineering work experience satisfactory to the Board that has been gained in addition to graduation, or any other education experience, and while performing engineering tasks under the direction of a person legally authorized to practice in an applicant's branch of engineering.

(b) An applicant for licensure as a land surveyor whose qualifications meet all requirements of the code and rules of the Board will be allowed to appear for only the second division of the written examination prescribed by Section 8741 of the Code if he or she meets one or more of the following requirements:

(1) Holds valid licensure as a professional civil engineer in California.

(2) Holds valid certification as an engineer-in-training obtained by passing a written examination which normally requires a minimum of eight hours to complete and the content of the examination is designed to test the candidate's knowledge of fundamental engineering subjects including mathematics and the basic sciences.

(3) Is a graduate of an approved land surveying curriculum and submits satisfactory evidence to the Board that he or she has fifteen (15) years or more of additional land surveying work experience satisfactory to the Board that has been gained in addition to graduation, or any other education experience, and while performing land surveying tasks under the direction of a person legally authorized to practice land surveying.

(4) Is a graduate of a nonapproved land surveying curriculum and submits satisfactory evidence to the Board that he or she has seventeen (17) years or more of additional land surveying work experience satisfactory to the Board that has been gained in addition to graduation, or any other education experience, and while performing land surveying tasks under the direction of a person legally authorized to practice in land surveying.

(5) Is a graduate of an approved land surveying curriculum and an approved postgraduate land surveying curriculum and submits satisfactory evidence to the Board that he or she

has fourteen (14) years or more of additional land surveying work experience satisfactory to the Board that has been gained in addition to graduation, or any other education experience, and while performing land surveying tasks under the direction of a person legally authorized to practice in land surveying.

(6) Is a graduate of a nonapproved land surveying curriculum and an approved post-graduate land surveying curriculum and submits satisfactory evidence to the Board that he or she has sixteen (16) years or more of additional land surveying work experience satisfactory to the Board that has been gained in addition to graduation, or any other education experience, and while performing land surveying tasks under the direction of a person legally authorized to practice in land surveying.

(7) Holds valid certification as a land surveyor-in-training in another state obtained by passing a written examination which normally requires a minimum of eight hours to complete and the content of the examination is designed to test the candidate's knowledge of fundamentals of land surveying including mathematics and the basic sciences.

(c) An applicant for a California certification as an engineer-in-training or a land surveyor-in-training who holds valid certification in another state obtained as in (a)(2) or (b)(7) above may be issued a California certificate.

439. Examination Not Permitted.

A person certified or licensed, by the board in a category or branch is not permitted to take any portion of the examination for the same category or branch unless his or her certificate or license has expired pursuant to Section 6796.3 or Section 8803 of the Code or unless reexamination is ordered as a probationary condition pursuant to Section 419.

441. Authorization to Take Examination.

(a) After evaluating the qualifications of an applicant and establishing that person's eligibility for the examination, in accordance with the applicable laws and regulations administered by the board, the executive officer assigns the applicant to the next scheduled examination for which the applicant qualified.

(b) Any applicant who lacks the qualifications for admission to the examination required by Chapter 7 or 15 of the code and rules of the board shall be declared ineligible; the application shall be denied and the application fee may be partially refunded in accordance with the provisions of Sections 158, 6763.5 and 8748.5 of the code. The executive officer shall notify each applicant of the reason for denying the application.

(c) Notification of the applicant's assignment and authorization to take the examination and the location shall be postmarked at least 14 days prior to the examination date.

442. Examination Subversion.

(a) Examination subversion is the use of any means to alter the results of an examination to cause the results to inaccurately represent the competency of an examinee. Examination subversion includes, but is not limited to:

(1) Communication between examinees inside of the examination room.

(2) Giving or receiving any unauthorized assistance on the examination while an examination is in progress.

(3) Having any unauthorized printed or written matter or other devices in his or her possession which might serve to aid the examinee on the examination.

(4) Obtaining, using, buying, selling, distributing, having possession of, or having unauthorized access to secured examination questions or other secured examination material prior to, during or after the administration of the examination.

(5) Copying another examinee's answers or looking at another examinee's materials while an examination is in progress.

(6) Permitting anyone to copy answers to the examination.

(7) Removing any secured examination materials from the examination facility.

(8) Allowing another person to take the examination in the examinee's place.

(9) Placing any identifying mark upon his or her examination papers other than his or her identification number or other identifiers as directed by the examination administrator.

(10) Use by an examinee of any written material, audio material, video material, digital material, or any other mechanism not specifically authorized during the examination for the purpose of assisting an examinee in the examination.

(11) Writing on anything other than designated examination material.

(12) Writing or erasing anything after time is called.

(b) At the discretion of the Executive Officer, if there is evidence of examination subversion by an examinee prior to, during, or after the administration of the examination, one or more of the following may occur:

(1) The examinee may be denied the privilege of taking the examination if examination subversion is detected before the administration of the examination.

(2) If the examination subversion detected has not yet compromised the integrity of the examination, such steps as are necessary to prevent further examination subversion shall be taken, and the examinee may be permitted to continue with the examination.

(3) The examinee may be requested to leave the examination facility if examination subversion is detected during the examination.

(4) The examinee may be requested to submit written advisement of his or her intent to comply with and understanding of the law.

(5) The examination results may be voided and the application fee forfeited.

(6) The examinee may not be allowed to sit for an examination for up to three (3) years.

(c) If examination subversion is detected after the administration of the examination, the Executive Officer shall make appropriate inquiry to determine the facts concerning the examination subversion and may take any of the actions as described in subdivision (b).

(d) The Executive Officer reserves the right not to release the examination results to the examinee pending the outcome of any investigation of examination subversion.

(e) Removal from or voidance of one part of a multiple-part examination taken during a single examination administration shall constitute removal from or voidance of all other parts of the multiple-part examination.

443. Inspection of Examination.

(a) The following definitions as used in these regulations have the meaning expressed in this section:

(1) "Essay type problem" means an engineering or land surveying problem in which the examinee provides a free response as solution and is graded for method, computations and answers. Scoring is determined by comparing examinee solutions with pre-determined scoring plans.

(2) "Multiple-choice type problem" means an engineering or land surveying problem which requires the examinee to select an answer from a multiple response format. Computations are not considered in the grading process for multiple-choice items.

(b) Multiple-choice type problems shall not be reviewed.

(c) An applicant who meets the criteria specified in subparagraph (1) or (2) below shall be granted 8 hours to review or write an appeal for an essay type problem or problems attempted during the written examination.

(1) Who is no more than 8 points below the passing score on a National Council of Examiners for Engineering and Surveying Professional Engineering examination;

(2) Who is no more than 15% below the passing score on a State specific Professional Engineering or Professional Land Surveying examination.

(d) Time and location of the examination review or appeal session shall be designated by the executive officer.

(e) At the time of the review or appeal session, no one other than the examinee and representatives of the board shall have access to such examination papers.

(f) Evidence that the applicant erased, deleted, removed or altered the examination papers, or the material contained thereon during such inspection, may result in the applicant being disqualified by the board from taking future examinations.

(g) The applicant shall have access to his or her examination, test booklets and solutions when reviewing or preparing an appeal of an essay type problem during the time period specified in subsection (c).

444. Examination Appeal.

(a) Applicants who were unsuccessful in the Professional Engineers or Professional Land Surveyor Examination will be notified by mail with the notice of results of the date and time in which they may attend a review or appeal session. The appeal fee referred to in section 407(d) must be received by the board within twenty-one (21) days from the date of the notice of the results of his or her examination. Applicants who decide to submit an appeal will be required to submit their appeal at the end of the review or appeal session.

(b) An appeal of an essay type problem of the examination shall be made in writing; and it shall state the reason for appeal, citing the item or items against which the appeal is directed, and it shall be accompanied by the appropriate appeal fee. The applicant shall identify the specific item(s) being challenged: the specific reasons for the challenge: and cite reference materials, facts and figures to substantiate the appeal. The appeal fee shall be payable for an appeal directed at an essay type problem of the examination. An appeal may be directed to any specific essay type problem or problems or sub-parts thereof, but an appeal shall cause the entire problem to be rescored. Rescoring of an essay type problem may result in one of the following three actions: points may be added; points may be deducted; or the score may remain the same. If the appeal results in the appellant being deemed to have passed the examination, the full appeal fee shall be refunded.

(c) The executive officer may deny any appeal requesting a review of an examination that is not accompanied by information supporting the reason for such request, is not accompanied by the correct appeal fee, or is not filed within the period of time provided in paragraph (a) of this rule.

(d) The Board's decision on an appeal of an essay type problem is final and shall not be re-evaluated. A multiple-choice type problem is not appealable.

446. Postponements.

The executive officer may grant a postponement, not to exceed two such postponements for each application to any applicant who for reasonable cause is prevented from appearing for examination at the time fixed, provided the applicant's request for postponement and the reason therefor is filed with the principal office of the board at any time prior to the expiration of the ten (10) day period immediately following the date of such examination.

447. Permissible Reference Material and Other Accessories.

The executive officer shall advise each examinee when he/she is notified that he/she is assigned to an examination, what reference and other materials may be used during the examination to which he/she has been assigned.

ARTICLE 4. MISCELLANEOUS

460. Curricula Approved by the Board.

(a) A curriculum approved by the board as qualifying a graduate of that curriculum for four years' engineering experience, or a non-graduate with one-half year of experience for each year of study completed, as provided in Section 6753 of the code, is defined as any engineering curriculum leading to a first degree in engineering accredited by the Accreditation Board for Engineering and Technology (ABET). The effective date of accreditation shall be one year prior to the initial year of accreditation as specified in the ABET Accreditation Yearbook.

(b) The board may give one-half year of experience credit for each year of study completed in a non-approved engineering curriculum except that the maximum of such experience credit shall be two years per applicant.

(c) (1) The board may give one-half year of experience credit for each year of study completed in an approved curriculum leading to a degree in engineering technology except that the maximum of such experience credit shall be two years per applicant.

(2) The board has approved the curricula leading to a degree in engineering technology which have been accredited by the Engineers' Council for Professional Development.

461. Testing Laboratory Reports.

Reports issued by Testing Laboratories shall be prepared by or under the supervision of a licensed civil, electrical, or mechanical engineer as appropriate and signed or sealed by him/her whenever such reports go beyond the tabulation of test data (composition of material, breaking stress, et cetera) and proceeds to:

(a) Interpret the data to draw conclusions as to the characteristics of civil engineering structure, an electrical or mechanical device, or parts thereof.

(b) Express civil, electrical, or mechanical engineering judgment in the form of recommendations derived from the results of the test.

(c) Perform design work in the preparation of plans, specifications, and other instruments requiring licensure as a civil, electrical, and/or mechanical engineer.

463. Notice of Association or Disassociation With Partnership, Firm, or Corporation.

(a) A professional engineer who is associated as the partner, member, officer, or employee in responsible charge of professional engineering services offered or performed by a firm, partnership, or corporation shall notify the Board within thirty (30) days of such association or termination of association on a form approved by the Board.

(b) A licensed land surveyor and/or civil engineer who practices or offers to practice land surveying, according to the provisions of Section 8729 of the Code, as a partner, member, or officer of a partnership, firm, or corporation shall advise the Board within thirty

(30) days of such association or termination of association on a form approved by the Board.

(c) A licensed photogrammetric surveyor who is associated as a member, partner, officer, or employee in a firm, partnership, or corporation which offers or performs photogrammetric surveying services according to the provisions of Article 5.5 of the Professional Land Surveyors' Act shall notify the Board within thirty (30) days of such association or termination of association on a form approved by the Board.

(d) A firm which contains partners, members or officers as described above will be allowed six months following the death, disassociation, or retirement of a member, partner, or officer whose name the firm, partnership, or corporation carried in its firm title to make a written request to the Board for an investigation pursuant to the requirements of Section 6738(e) of the Code. The Board will determine if such firm is eligible to continue use of its firm title without change.

463.5 Providing Notice of Licensure.

Every licensee shall provide notice to his or her clients that the licensee is licensed by the Board for Professional Engineers and Land Surveyors. Notice shall be provided by one or more of the following methods:

(a) Displaying his or her wall certificate in a public area; or office; or individual work area of the premises where the licensee provides the licensed service.

(b) Providing a statement to each client that states the client understands the licensee is licensed by the Board for Professional Engineers and Land Surveyors. Said statement shall be signed and dated by the client and shall be retained in the licensee's records.

(c) Including a statement that the licensee is licensed by the Board for Professional Engineers and Land Surveyors either on letterhead or on a contract for services. If said statement is included on a contract for services, it shall be placed immediately above the signature line for the client in at least 12 point type.

(d) Posting a notice in a public area of the premises where the licensee provides the licensed services that states the named licensee is licensed by the Board for Professional Engineers and Land Surveyors. Said notice shall be in at least 48 point type.

464. Corner Record.

(a) The corner record required by Section 8773 of the Code for the perpetuation of monuments shall contain the following information for each corner identified therein:

(1) The county and, if applicable, city in which the corner is located.

(2) An identification of the township, range, base, and meridian in which the corner is located, if applicable.

(3) Identification of the corner type (example: government corner, control corner, property corner, etc.).

(4) Description of the physical condition of

(A) the monument as found and

(B) any monuments set or reset.

(5) The date of the visit to the monument when the information for the corner record was obtained.

(6) For Public Land Corners for which a corner record is required by Section 8773(a) of the Code, a sketch shall be made showing site recovery information that was used for the corner. For other kinds of corners, a drawing shall be made which shows measurements that relate the corner to other identifiable monuments.

(7) A reference to the California Coordinate System is optional at the discretion of the preparer of the record.

(8) The date of preparation of the corner record and, as prescribed by Section 8773.4 of the Code, the signature and title of the chief of the survey party if the corner record is prepared by a United States Government or California State agency or the signature and seal of the land surveyor or civil engineer, as defined in Section 8731 of the Code, preparing the corner record.

(9) The date the corner record was filed and the signature of the county surveyor.

(10) A document or filing number.

(b) A corner record shall be filed for each public land survey corner which is found, reset, or used as control in any survey by a land surveyor or a civil engineer. Exceptions to this rule are identified in Section 8773.4 of the Code.

(c) The corner record shall be filed within 90 days from the date a corner was found, set, reset, or used as control in any survey. The provisions for extending the time limit shall be the same as provided for a record of survey in Section 8762 of the Code.

(d) A corner record may be filed for any property corner, property controlling corner, reference monument, or accessory to a property corner, together with reference to record information. Such corner record may show one or more property corners, property controlling corners, reference monuments, or accessories to property corners on a single corner record document so long as it is legible, clear, and understandable.

(e) When conducting a survey which is a retracement of lines shown on a subdivision map, official map, or a record of survey, where no material discrepancies with these records are found and where sufficient monumentation is found to establish the precise location of property corners thereon, a corner record may be filed in lieu of a record of survey for any property corners which are set or reset or found to be of a different character than indicated by prior records. Such corner records may show one or more property corners, property controlling corners, reference monuments or accessories to property corners on a single corner record document so long as it is legible, clear, and understandable.

(f) The standard markings and standard abbreviations used by the Bureau of Land Management (formerly the General Land Office) of the United States Department of the Interior shall be used in the corner record.

(g) The corner record shall be filed on a form prescribed by the Board. The approved form is BORPELS-1297.

465. Records of Survey—Public Officers.

A public officer who has performed a survey which is subject to the requirements of Section 8762 of the code, shall file a record of survey map or comply with Section 8765(a) of the code. In either event, the public officer shall file the information required by Section 8765(a) of the code relating to his or her survey within the time limit provided for in Section 8762 of the code.

470. Application Response.

(a) The following timeframes, based on taking and passing the first available examination, shall apply to applications for licensure as a professional engineer or as a professional land surveyor when an examination is required.

(1) Within 150 calendar days of receipt of an application, the Board shall inform the applicant in writing that the application is either accepted for filing or that it is deficient and what specific information or documentation is required to complete the application.

(2) A complete application shall include a score for all parts of the examination. Within 60 calendar days after the filing of a complete application, the Board shall make a decision regarding the applicant's eligibility for licensure.

(b) The following timeframes, based on taking and passing the first available examination, shall apply to applications for certification to use the titles "Engineer-in-Training" or "Land Surveyor-in-Training" when an examination is required:

(1) Within 75 calendar days of receipt of an application, the Board shall inform the applicant in writing that the application is either accepted for filing or that it is deficient and what specific information or documentation is required to complete the application.

(2) A complete application shall include a score for the examination. Within 60 calendar days after the filing of a complete application, the Board shall make a decision regarding the applicant's eligibility to use either the title "Engineer-in-Training" or "Land Surveyor-in-Training."

(c) The following timeframes shall apply to applications for licensure as a professional engineer or as a professional land surveyor when comity licensure is applied for and when no examination is required:

(1) Within 125 calendar days of receipt of a comity application, the Board shall inform the applicant in writing that the application is either accepted for filing and complete or that it is deficient and what specific information or documentation is required to complete the application.

(2) Within 60 calendar days after the date of filing of a complete application, the Board shall

make a decision regarding the application for licensure.

471. Processing Time.

The Board's time periods are based upon an applicant taking and passing the first available examination and for processing an application from the receipt of the initial application to the final decision regarding issuance or denial of licensure based on the board's actual performance during the two years preceding the proposal of this section:

(a) The median time was 255 days.

(b) The minimum time was 10 days.

(c) The maximum time was 1548 days.

472. Citations of Unlicensed Persons.

(a) The Executive Officer or his or her designee may issue a citation for any violation of any provision of law enforced by the Board to an unlicensed person who, unless otherwise exempt, is acting in the capacity of a professional engineer or professional land surveyor .

(b) Each citation

(1) shall be in writing;

(2) shall describe with particularity the nature of the violation, including specific reference to the provision or provisions of law determined to have been violated;

(3) shall contain an assessment of an administrative fine, an order of abatement fixing a reasonable period of time for abatement of the violation, or both an administrative fine and an order of abatement;

(4) shall be served on the cited person at the last known business or residence address personally or by certified mail with return receipt requested;

(5) shall inform the cited person that failure to pay the fine within 30 calendar days of the date of assessment, unless the citation is being appealed, may result in the Executive Officer applying to the appropriate superior court for a judgment in the amount of the administrative fine;

(6) shall inform the cited person that, if he/she desires an informal conference to contest the finding of a violation, the informal conference shall be requested by written notice to the Board within 30 calendar days from service of the citation;

(7) shall inform the cited person that, if he or she desires a hearing to contest the finding of a violation, that hearing shall be requested by written notice to the Board within 30 calendar days from service of the citation.

472.1. Assessment of Administrative Fine.

(a) In no event shall the administrative fine be less than $50 or exceed $5,000 for each violation.

(b) In determining the amount of an administrative fine , the Executive Officer shall consider the following factors:

(1) The nature and severity of the violation;

(2) The good or bad faith exhibited by the cited person;

(3) The history of previous violations;

(4) The extent to which the cited person has cooperated with the Board and the Board's investigation;

(5) The extent to which the cited person has mitigated or attempted to mitigate any damage or injury caused by his/her violation;

(6) Any factors in extenuation or aggravation related to the violation;

(7) Other matters as may be appropriate.

472.2. Appeal of Citations.

(a) Any person served with a citation issued pursuant to Section 472 may contest the citation by submitting a written request for a hearing to the Board within 30 calendar days of service of the citation. Such hearings shall be conducted pursuant to the Administrative Procedure Act, Chapters 4.5 and 5, commencing with Section 11400) of Part 1 of Division 3 of Title 2 of the Government Code.

(b) In addition to requesting a hearing as described in subsection (a), the cited person may, within 30 calendar days of service of the citation, submit a written request for an informal conference with the Executive Officer.

(c) The Executive Officer may, within 30 working days from receipt of a written request for an informal conference as provided in subsection (b), hold an informal conference with the cited person and/or his or her legal counsel or authorized representative. The 30-day period may be extended by the Executive Officer for good cause. Following the informal conference, the Executive Officer may affirm, modify, or dismiss the citation, including any administrative fine assessed or order of abatement issued. An order affirming, modifying, or dismissing the original citation shall be served on the cited person within 30 calendar days from the informal conference. Said order shall state in writing the reasons for the affirmation, modification, or dismissal of the original citation. If the order affirms or modifies the original citation, said order shall fix a reasonable period of time for abatement of the violation or payment of the fine. Service of this order shall be made as provided in Section 472. This order shall be considered the conclusion of the informal conference proceedings.

(d) If the citation is affirmed or modified following the informal conference, the cited person may request a hearing as provided in subsection (a) within 30 calendar days from service of the order described in subsection (c). The cited person shall not be permitted to request another informal conference.

(e) If the citation is dismissed after the informal conference, the request for a hearing, if any, shall be deemed to be withdrawn.

(f) Submittal of a written request for a hearing as provided in subsection (a), an informal conference as provided in subsection (b), or both stays the time period in which to abate the violation and/or to pay the fine.

(g) If the written request for a hearing as provided in subsection (a) or an informal conference as provided in subsection (b) or both is not submitted within 30 calendar days from service of the citation, the cited person is deemed to have waived his/her right to a hearing or an informal conference.

472.3. Compliance with Order.

(a) If a cited person who has been issued an order of abatement is unable to complete the correction within the time set forth in the citation because of conditions beyond his/her control after the exercise of reasonable diligence, then he/she may request from the Executive Officer an extension of time within which to complete the correction. Such a request shall be in writing and shall be made within the time set forth for abatement.

(b) Failure of an applicant for licensure as a professional engineer or as a professional land surveyor to abate the violation or to pay the fine within the time allowed is a ground for denial of licensure.

(c) Notwithstanding any other provision of the law, the Executive Officer may waive all or part of an administrative fine if the person against whom the citation is assessed satisfactorily completes all the requirements for, and is issued, a license.

(d) If the cited person fails to pay the assessed fine within the time allowed, the Executive Officer may apply to the appropriate superior court for a judgment in the amount of the administrative fine.

(e) If a hearing as provided in Section 472.2(a) is not requested, payment of the fine shall not constitute an admission of the violation charged.

(f) Payment of the fine shall be represented as satisfactory resolution of the matter for purposes of public disclosure.

472.4. Disconnection of Telephone Service.

(a) If, upon investigation, the Executive Officer has cause to believe that an unlicensed individual acting in the capacity of a professional engineer or professional land surveyor, who is not otherwise exempted from the provisions of the Professional Engineers Act or the Professional Land Surveyors' Act, has violated Sections 6787 or 8792 of the Code by advertising in a telephone directory, without being properly licensed, the Executive Officer may issue a citation under Section 472 containing an order of abatement which requires the cited person to both cease the unlawful advertising and notify the telephone company furnishing services to the cited person to disconnect the telephone services furnished to any telephone number contained in the unlawful advertising, and that subsequent calls to that number shall not be referred by the telephone company to any new telephone number obtained by that person. The cited person shall provide written evidence of compliance to the Executive Officer.

(b) If the cited person fails to comply with an order of abatement as described in subsection (a) within the time allowed, the Executive Officer shall inform the Public Utilities Commission of the violation in accordance with Business and Professions Code section 149.

473. Citations of Licensed Persons.

(a) The Executive Officer or his or her designee may issue a citation for any violation of any provision of law enforced by the Board to any person who holds a license issued by the Board.

(b) Prior to the issuance of a citation in any case involving allegations of negligence and/or incompetence, as defined in Section 404, in the professional practice, the Executive Officer shall submit the alleged violation for review to at least one licensee of the Board who is competent in the branch of professional engineering or professional land surveying most relevant to the subject matter of the citation. The licensee reviewing the alleged violation shall be either a member of the Board's professional staff, a technical advisory committee member, or an expert consultant. Upon conclusion of the review, the reviewer shall prepare a finding of fact and a recommendation based upon that finding to which the Executive Officer shall give due consideration in determining whether cause exists to issue a citation.

(c) Each citation

    (1) shall be in writing;

    (2) shall describe with particularity the nature of the violation, including specific reference to the provision or provisions of law determined to have been violated;

    (3) may contain an assessment of an administrative fine, an order of abatement fixing a reasonable period of time for abatement of the violation, or both an administrative fine and an order of abatement;

    (4) shall be served on the cited person at the address of record on file with the Board

personally or by certified mail with return receipt requested;

(5) shall inform the cited person that failure to pay the fine within 30 calendar days of the date of assessment, unless the citation is being appealed, may result in disciplinary action being taken by the Board. If a citation is not contested and the fine is not paid, the full amount of the assessed fine shall be added to the fee for renewal of the license. A license shall not be renewed without the payment of the renewal fee and fine;

(6) shall inform the cited person that, if he/she desires an informal conference to contest the finding of a violation, the informal conference shall be requested by written notice to the Board within 30 calendar days from service of the citation;

(7) shall inform the cited person that, if he or she desires a hearing to contest the finding of a violation, that hearing shall be requested by written notice to the Board within 30 calendar days from service of the citation.

{Amended, effective September 24, 2005}

473.1. Assessment of Administrative Fines.

(a) In no event shall the administrative fine be less than $50 or exceed $5,000 for each violation.

(b) In determining the amount of an administrative fine, the Executive Officer shall consider the following factors:

(1) The nature and severity of the violation;

(2) The good or bad faith exhibited by the cited person;

(3) The history of previous violations;

(4) The extent to which the cited person has cooperated with the Board and the Board's investigation;

(5) The extent to which the cited person has mitigated or attempted to mitigate any damage or injury caused by his/her violation;

(6) Any factors in extenuation or aggravation related to the violation;

(7) Other matters as may be appropriate.

473.2. Appeal of Citations.

(a) Any person served with a citation issued pursuant to Section 473 may contest the citation by submitting a written request for a hearing to the Board within 30 calendar days of service of the citation. Such hearings shall be conducted pursuant to the Administrative Procedure Act, Chapters 4.5 and 5, commencing with Section 11400) of Part 1 of Division 3 of Title 2 of the Government Code.

(b) In addition to requesting a hearing as described in subsection (a), the cited person may, within 30 calendar days of service of the citation, submit a written request for an informal conference with the Executive Officer.

(c) The Executive Officer shall, within 30 working days of receipt of a written request for an informal conference as provided in subsection (b), hold an informal conference with the cited person and/or his or her legal counsel or authorized representative. The 30-day period may be extended by the Executive Officer for good cause. Following the informal conference, the Executive Officer may affirm, modify, or dismiss the citation, including any administrative fine assessed or order of abatement issued. An order affirming, modifying, or dismissing the original citation shall be served on the cited person within 30 calendar days from the informal conference. Said order shall state in writing the reasons for the affirmation, modification, or dismissal of the original citation. If the order affirms or modifies the original citation, said order shall fix a reasonable period of time for abatement of the violation or payment of the fine. Service of this order shall be made as provided in Section 473. This order shall be considered the conclusion of the informal conference proceedings.

(d) If the citation is affirmed or modified following the informal conference, the cited person may request a hearing as provided in subsection (a) within 30 calendar days from service of the order described in subsection (c). The cited person shall not be permitted to request another informal conference. ,

(e) If the citation is dismissed after the informal conference, the request for a hearing, if any, shall be deemed to be withdrawn.

(f) Submittal of a written request for a hearing as provided in subsection (a), an informal conference as provided in subsection (b), or both stays the time period in which to abate the violation and/or to pay the fine.

(g) If the written request for a hearing as provided in subsection (a) or an informal conference as provided in subsection (b) or both is not submitted within 30 calendar days from service of the citation, the cited person is deemed to have waived his/her right to a hearing or an informal conference.

473.3. Compliance with Citations.

(a) If a cited person who has been issued an order of abatement is unable to complete the correction within the time set forth in the citation because of conditions beyond his/her control after the exercise of reasonable diligence, then he/she may request from the Executive Officer an extension of time within which to complete the correction. Such a request shall be in writing and shall be made within the time set forth for abatement.

(b) Failure of the cited person to abate the violation or to pay the fine within the time allowed is grounds for suspension or revocation of the cited person's license.

(c) If the cited person fails to pay all of the fine within the time allowed, the balance due for the fine shall be added to the renewal fee for the license, and the license shall not be renewed until the fine is paid in full.

(d) If a hearing as provided in Section 473.2(a) is not requested, payment of the fine shall not constitute an admission of the violation charged.

(e) Payment of the fine shall be represented as satisfactory resolution of the matter for purposes of public disclosure.

473.4. Disconnection of Telephone Service.

(a) If, upon investigation, the Executive Officer has cause to believe that a professional engineer or professional land surveyor, who is not otherwise exempted from the provisions of the Professional Engineers Act or the Professional Land Surveyors' Act, has violated Sections 6787 or 8792 of the Code by advertising in a telephone directory, without being properly licensed, the Executive Officer may issue a citation under Section 473 containing an order of abatement which requires the cited person to both cease the unlawful advertising and notify the telephone company furnishing services to the cited person to disconnect the telephone services furnished to any telephone number contained in the unlawful advertising, and that subsequent calls to that number shall not be referred by the telephone company to any new telephone number obtained by that person. The cited person shall provide written evidence of compliance to the Executive Officer.

(b) If the cited person fails to comply with an order of abatement as described in subsection (a) within the time allowed, the Executive Officer shall inform the Public Utilities Commission of the violation in accordance with Business and Professions Code section 149.

474. Establishment of Criteria.

The criteria for the selection of a contractor are: professional excellence, demonstrated competence, specialized experience of the person, education and experience, ability to meet schedules, nature and quality of completed work, reliability of the person, location and other considerations the Executive Officer deems necessary to the performance of the contract.

474.1. Request for Qualifications.

(a) Where a project requires professional engineering or land surveying services the Executive Officer shall make a statewide request for qualifications through the publications of the respective professional societies and in other appropriate publications.

(b) The request for qualifications shall contain the following information: the nature of the work, the criteria upon which the award shall be made, the name of the contact person, the address to send statements of qualifications and performance data , and the deadline by which the statements must be received.

(c) The Executive Officer shall provide a copy of each request for qualifications to all small businesses who have indicated an interest in receiving the request. A failure of the Executive Officer to send a copy of a request for qualifications to any person shall not operate to preclude any contract.

474.2. Selection of Engineers and Land Surveyors.

After expiration of the deadline stated in the publications, as specified in Section 474.1, the Executive Officer, in conjunction with no less than three persons selected by the Executive Officer, shall evaluate statements of qualifications and performance data of firms which have been submitted to the Board. Based upon the statement of qualifications and performance data, a list ranking the qualified firms will be made.

From a list of three or more qualified firms, the Executive Officer shall select the first three ranked firms for contract negotiation to provide the services required. Where three qualified firms cannot be found which could provide the required service, the Executive Officer may then select from the available firms or issue a new request for proposal. Selections based on less than three firms must be documented with the names and addresses of firms contacted by the Executive Officer and the reasons why the provisions of this section could not be met. This documentation shall be maintained in the Boards Contract file.

474.3. Conflict of Interest/Unlawful Activity.

Governmental agency employees who are related to persons seeking to contract under these regulations shall not participate in any aspect of the contract review or selection process.

Further, any practice that results in an unlawful activity including, but not limited to, rebates, kickbacks, or any other unlawful consideration shall be prohibited.

474.4. Amendments.

In instances where the Board effects a necessary change in the project during the course of performance of the contract, the contractor's compensation may be adjusted by negotiation of a mutual written agreement in a fair and reasonable amount where the amount of work to be performed by the contractor is changed from that which the parties had originally contemplated.

474.5. Contracting in Phases.

Should the Board determine that it is necessary or desirable to have a given project performed in phases, it will not be necessary to negotiate the total contract price or compensation provisions in the initial instance, provided that the Board shall have determined that the person is best qualified to perform the whole project at a fair and reasonable cost, and the contract contains provisions that the Board, at it option, may utilize the person for other phases and that the person will accept a fair and reasonable price for subsequent phases to be later negotiated and reflected in a subsequent written instrument. The procedure with regard to negotiation provided for in Section 6106 of the Public Contract Code shall be applicable.

475. Code of Professional Conduct—Professional Engineering.

To protect and safeguard the health, safety, welfare, and property of the public, every person who is licensed by the Board as a professional engineer, including licensees employed in any manner by a governmental entity or in private practice, shall comply with this Code of Professional Conduct. A violation of this Code of Professional Conduct in the practice of professional engineering constitutes unprofessional conduct and is grounds for disciplinary action pursuant to Section 6775 of the Code. This Code of Professional Conduct shall be used for the sole purpose of investigating complaints and making findings thereon under Section 6775 of the Code.

(a) Compliance with Laws Applicable to a Project:

A licensee shall provide professional services for a project in a manner that is consistent with the laws, codes, ordinances, rules, and regulations applicable to that project. A licensee may obtain and rely upon the advice of other professionals (e.g., architects, attorneys, professional engineers, professional land surveyors, and other qualified persons) as to the intent and meaning of such laws, codes, and regulations.

(b) Conflict of Interest:

(1) If a licensee provides professional services for two or more clients on a project or related projects, the licensee shall disclose in writing to those clients and property owners or their

authorized representatives his or her relationship to those clients.

(2) If a licensee has a business association or a financial interest which may influence his or her judgment in connection with the performance of professional services, the licensee shall fully disclose in writing to his or her client(s) or employer(s) the nature of the business association or the financial interest.

(3) A licensee shall not solicit or accept payments, rebates, refunds, or commissions, whether in the form of money or otherwise, from contractors or suppliers of material, systems, or equipment in return for specifying their products to a client or employer of the licensee.

(4) A licensee, while engaged by a governmental agency as an officer, employee, appointee, agent, or consultant of that agency shall not engage in a professional engineering business or activity that may be subject to that licensee's direct or indirect control, inspection, review, audit, or enforcement on behalf of that agency, unless the circumstances are disclosed to and approved by that agency in writing prior to such engagement.

(c) Representations:

(1) A licensee shall not misrepresent his or her qualifications to a prospective or existing client or employer.

(2) A licensee shall not misrepresent to a prospective or existing client the licensee's scope of responsibility in connection with projects or services for which the licensee is receiving or will receive compensation from that client.

(3) A licensee shall not misrepresent his or her scope of responsibility in connection with projects or services for which the licensee is claiming credit.

(4) A licensee shall not misrepresent nor permit the misrepresentation of his or her professional qualifications, or affiliations or the affiliations or purposes of the institutions, organizations, or other businesses with which he or she is associated.

(5) When providing information in connection with a person's application for a license to practice professional engineering, a licensee shall accurately represent his or her knowledge of the applicant's qualifications.

(6) A licensee may advertise or solicit for any services for which he or she is authorized by licensure.

(7) A licensee shall only express professional opinions that have a basis in fact or experience or accepted engineering principles.

(8) A licensee shall attribute proper credit to others for their professional work or professional contribution and shall not misappropriate the professional work of others.

(9) A licensee shall not knowingly permit the publication or use of his or her data, reports, plans, or other professional documents for unlawful purposes.

(10) A licensee shall not falsely or maliciously injure or attempt to injure the reputation or business of others.

(11) A licensee shall not misrepresent data and/or its relative significance in any professional engineering report.

(d) Confidential Information:

Confidential information obtained by a licensee, in his or her professional capacity, concerning a client, employer, or other related party shall not be disclosed by the licensee without the permission of the client, employer, or other related party except for the following:

(1) Disclosures made in response to an order of the court or to a subpoena or summons enforceable by an order of the court.

(2) Disclosures made in an adjudicatory proceeding.

(3) Disclosures made in response to an official inquiry from a governmental regulatory agency.

(4) Disclosures made when required by law.

(5) Disclosures made upon discovering a hazard within the licensee's field of professional expertise which may threaten the health, safety, and welfare of the public.

(6) Disclosures made when providing evidence to the Board regarding other licensees or unlicensed individuals who may have violated the Professional Engineers Act.

(7) Disclosures made regarding illegal conduct.

As used in this section, "confidential information" means information identified as confidential by the licensee's client, employer, or other related party.

(e) Document Submittal:

(1) A licensee shall not misrepresent the completeness of the professional documents he or she submits to a governmental agency.

(2) A licensee shall not misrepresent the completeness of the professional documents he or she prepared to his or her client or to other involved parties.

476. Code of Professional Conduct—Professional Land Surveying.

To protect and safeguard the health, safety, welfare, and property of the public, every person who is licensed by the Board as a professional land surveyor or professional civil engineer legally authorized to practice land surveying, including licensees employed in any manner by a governmental entity or in private practice, shall comply with this Code of Professional Conduct. A violation of this Code of Professional Conduct in the practice of professional land surveying constitutes unprofessional conduct and is grounds for disciplinary action pursuant to Section 8780 of the Code. This Code of Professional Conduct shall be used for the sole purpose of investigating complaints and making findings thereon under Section 8780 of the Code.

(a) Compliance with Laws Applicable to a Project:

A licensee shall provide professional services for a project in a manner that is consistent with the laws, codes, ordinances, rules, and regulations applicable to that project. A licensee may obtain and rely upon the advice of other professionals (e.g., architects, attorneys, professional engineers, professional land surveyors, and other qualified persons) as to the intent and meaning of such laws, codes, and regulations.

(b) Conflict of Interest:

(1) If a licensee provides professional services for two or more clients on a project or related projects, the licensee shall disclose in writing to those clients and property owners or their authorized representatives his or her relationship to those clients.

(2) If a licensee has a business association or a financial interest which may influence his or her judgment in connection with the performance of professional services, the licensee shall fully disclose in writing to his or her client(s) or employer(s) the nature of the business association or the financial interest.

(3) A licensee shall not solicit or accept payments, rebates, refunds, or commissions, whether in the form of money or otherwise, from contractors or suppliers of material, systems, or equipment in return for specifying their products to a client or employer of the licensee.

(4) A licensee, while engaged by a governmental agency as an officer, employee, appointee, agent, or consultant of that agency shall not engage in a professional land surveying business or activity that may be subject to that licensee's direct or indirect control, inspection, review, audit, or enforcement on behalf of that agency, unless the circumstances are disclosed to and approved by that agency in writing prior to such engagement.

(c) Representations:

(1) A licensee shall not misrepresent his or her qualifications to a prospective or existing client or employer.

(2) A licensee shall not misrepresent to a prospective or existing client the licensee's scope of responsibility in connection with projects or services for which the licensee is receiving or will receive compensation from that client.

(3) A licensee shall not misrepresent his or her scope of responsibility in connection with projects or services for which the licensee is claiming credit.

(4) A licensee shall not misrepresent nor permit the misrepresentation of his or her professional qualifications, or affiliations or the affiliations or purposes of the institutions, organizations, or other businesses with which he or she is associated.

(5) When providing information in connection with a person's application for a license to practice professional land surveying, a licensee

shall accurately represent his or her knowledge of the applicant's qualifications.

(6) A licensee may advertise or solicit for any services for which he or she is authorized by licensure.

(7) A licensee shall only express professional opinions that have a basis in fact or experience or accepted land surveying principles.

(8) A licensee shall attribute proper credit to others for their professional work or professional contribution and shall not misappropriate the professional work of others.

(9) A licensee shall not knowingly permit the publication or use of his or her data, reports, maps, or other professional documents for unlawful purposes.

(10) A licensee shall not falsely or maliciously injure or attempt to injure the reputation or business of others.

(11) A licensee shall not misrepresent data and/or its relative significance in any professional land surveying report.

(d) Confidential Information:

Confidential information obtained by a licensee, in his or her professional capacity, concerning a client, employer, or other related party shall not be disclosed by the licensee without the permission of the client, employer, or other related party except for the following:

(1) Disclosures made in response to an order of the court or to a subpoena or summons enforceable by an order of the court.

(2) Disclosures made in an adjudicatory proceeding.

(3) Disclosures made in response to an official inquiry from a governmental regulatory agency.

(4) Disclosures made when required by law.

(5) Disclosures made upon discovering a hazard within the licensee's field of professional expertise which may threaten the health, safety, and welfare of the public.

(6) Disclosures made when providing evidence to the Board regarding other licensees or unlicensed individuals who may have violated the Professional Land Surveyors' Act.

(7) Disclosures made regarding illegal conduct.

As used in this section, "confidential information" means information identified as confidential by the licensee's client, employer, or other related party.

(e) Document Submittal:

(1) A licensee shall not misrepresent the completeness of the professional documents he or she submits to a governmental agency.

(2) A licensee shall not misrepresent the completeness of the professional documents he or she prepared to his or her client or to other involved parties.

Professional Engineer's Act, 2007

Business and Professions Code §§ 6700–6799

INCLUDES AMENDMENTS MADE DURING THE 2007 LEGISLATIVE SESSION
(Effective January 1, 2008, unless otherwise noted)

CHAPTER 7. PROFESSIONAL ENGINEERS
ARTICLE 1. GENERAL PROVISIONS

6700. Professional Engineers Act

This chapter constitutes the chapter on professional engineers. It may be cited as the Professional Engineers Act.

6701. Professional Engineer Defined

"Professional engineer," within the meaning and intent of this act, refers to a person engaged in the professional practice of rendering service or creative work requiring education, training and experience in engineering sciences and the application of special knowledge of the mathematical, physical and engineering sciences in such professional or creative work as consultation, investigation, evaluation, planning or design of public or private utilities, structures, machines, processes, circuits, buildings, equipment or projects, and supervision of construction for the purpose of securing compliance with specifications and design for any such work.

6702. Civil Engineer Defined

"Civil engineer" as used in this chapter means a professional engineer in the branch of civil engineering and refers to one who practices or offers to practice civil engineering in any of its phases.

6702.1. Electrical Engineer Defined

"Electrical engineer" as used in this chapter means a professional engineer in the branch of electrical engineering and refers to one who practices or offers to practice electrical engineering in any of its phases.

6702.2. Mechanical Engineer Defined

"Mechanical engineer" as used in this chapter means a professional engineer in the branch of mechanical engineering and refers to one who practices or offers to practice mechanical engineering in any of its phases.

6703. Responsible Charge of Work Defined

The phrase "responsible charge of work" means the independent control and direction, by the use of initiative, skill, and independent judgment, of the investigation or design of professional engineering work or the direct engineering control of such projects. The phrase does not refer to the concept of financial liability.

6703.1. Supervision of Construction Defined

"Supervision of the construction of engineering structures" means the periodic observation of materials and completed work to determine general compliance with plans, specifications, and design and planning concepts. However, "supervision of construction of engineering structures" does not include responsibility for the superintendence of construction processes, site conditions, operations, equipment, personnel, or the maintenance of a safe place to work or any safety in, on, or about the site.

For purposes of this subdivision, "periodic observation" means visits by an engineer, or his or her agent, to the site of a work of improvement.

6704. Defines Who May Use Engineer Titles

(a) In order to safeguard life, health, property, and public welfare, no person shall practice civil, electrical, or mechanical engineering unless appropriately licensed or specifically exempted from licensure under this chapter, and only persons licensed under this chapter shall be entitled to take and use the titles "consulting engineer," "professional engineer," or "registered engineer," or any combination of those titles or abbreviations thereof, and according to licensure with the board the engineering branch titles specified in Section 6732, or the authority titles specified in Sections 6736 and 6736.1, or the title "engineer-in-training."

(b) The provisions of this section shall not prevent the use of the title "consulting engineer" by a person who has qualified for and maintained exemption for using that title under the provisions of Section 6732.1, or by a person licensed as a photogrammetric surveyor.

6704.1. Title Act Review

(a) The Department of Consumer Affairs, in conjunction with the board, and the Joint Committee on Boards, Commissions, and Consumer Protection shall review the engineering branch titles specified in Section 6732 to determine whether certain title acts should be eliminated from this chapter, retained, or converted to practice acts similar to civil, electrical, and mechanical engineering, and whether supplemental engineering work should be permitted for all branches of engineering. The department shall contract with an independent consulting firm to perform this comprehensive analysis of title act registration.

(b) The independent consultant shall perform, but not be limited to, the following:

(1) meet with representatives of each of the engineering branches and other professional groups;

(2) examine the type of services and work provided by engineers in all branches of engineering and interrelated professions within the marketplace, to determine the interrelationship that exists between the various branches of engineers and other interrelated professions;

(3) review and analyze educational requirements of engineers;

(4) identify the degree to which supplemental or "overlapping" work between engineering branches and interrelated professions occurs;

(5) review alternative methods of regulation of engineers in other states and what impact the regulations would have if adopted in California;

(6) identify the manner in which local and state agencies utilize regulations and statutes to regulate engineering work; and,

(7) recommend changes to existing laws regulating engineers after considering how these changes may effect the health, safety, and welfare of the public.

(c) The board shall reimburse the department for costs associated with this comprehensive analysis. The department shall report its findings and recommendations to the Legislature by September 1, 2002.

6705. Subordinate Defined

A subordinate is any person who assists a registered professional engineer in the practice of professional engineering without assuming responsible charge of work.

6706. Good Samaritan Immunity

(a) An engineer who voluntarily, without compensation or expectation of compensation, provides structural inspection services at the scene of a declared national, state, or local emergency at the request of a public official, public safety officer, or city or county building inspector acting in an official capacity shall not be liable in negligence for any personal injury, wrongful death, or property damage caused by the engineer's good faith but negligent inspection of a structure used for human habitation or owned by a public entity for structural integrity or nonstructural elements affecting life and safety.

The immunity provided by this section shall apply only for an inspection that occurs within 30 days of the declared emergency. Nothing in this section shall provide immunity for gross negligence or willful misconduct.

(b) As used in this section:

(1) "Engineer" means a person registered under this chapter as a professional engineer, including any of the branches thereof.

(2) "Public safety officer" has the meaning given in Section 3301 of the Government Code.

(3) "Public official" means a state or local elected officer.

6706.3. References to Registered Engineer Deemed to Refer to Licensed Engineer

Any reference in any law or regulation to a registered engineer, or to a registered civil, electrical, or mechanical engineer, is deemed to refer to a licensed engineer, or to a licensed civil, electrical, or mechanical engineer, as the case may be.

ARTICLE 2. ADMINISTRATION

6710. Board Name and Composition; Reference to Previous Name; Sunset Dates

(a) There is in the Department of Consumer Affairs a Board for Professional Engineers and Land Surveyors, which consists of 13 members.

(b) Any reference in any law or regulation to the Board of Registration for Professional Engineers and Land Surveyors is deemed to refer to the Board for Professional Engineers and Land Surveyors.

(c) This section shall become inoperative on July 1, 2011, and, as of January 1, 2012, is repealed, unless a later enacted statute, that becomes effective on or before January 1, 2012, deletes or extends the dates on which it becomes inoperative and is repealed. The repeal of this section renders the board subject to the review required by Division 1.2 (commencing with Section 473).

6710.1 Legislative Intent—Protection of the Public

Protection of the public shall be the highest priority for the Board for Professional Engineers and Land Surveyors in exercising its licensing, regulatory, and disciplinary functions. Whenever protection of the public is inconsistent with other interests sought to be promoted, the protection of the public shall be paramount.

6711. Qualifications of Board Members

Each member of the board shall be a citizen of the United States. Five members shall be registered under this chapter. One member shall be licensed under the Land Surveyors' Act, Chapter 15 (commencing with Section 8700) of this division, and seven shall be public members who are not registered under this act or licensed under the Land Surveyors' Act. Each member, except the public members, shall have at least 12 years active experience and shall be of good standing in his profession. Each member shall be at least 30 years of age, and shall have been a resident of this state for at least five years immediately proceeding his appointment.

6712. Appointments; Term; Qualifications

(a) All appointments to the board shall be for a term of four years. Vacancies shall be filled by appointment for the unexpired term. Each appointment thereafter shall be for a four-year term expiring on June 30 of the fourth year following the year in which the previous term expired.

(b) Each member shall hold office until the appointment and qualification of his or her successor or until one year shall have elapsed since the expiration of the term for which he or she was appointed, whichever first occurs. No person shall serve as a member of the board for more than two consecutive terms.

(c) The Governor shall appoint professional members so that one is licensed to practice engineering as a civil engineer, one as an electrical engineer, one as a mechanical engineer, another is authorized to use the title of structural engineer, and one is a member of one of the remaining branches of engineering. One of the professional members licensed under this chapter or under Chapter 15 (commencing with Section 8700) shall be from a local public agency, and one shall be from a state agency.

(d) The Governor shall appoint five of the public members and the professional members qualified as provided in Section 6711. The Senate Rules Committee and the Speaker of the Assembly shall each appoint a public member.

{Amended by Stats.2007, Ch. 354}

6713. Removal of Members

The Governor may remove any member of the board for misconduct, incompetency or neglect of duty.

6714. Appointment of Executive Officer; Salary

The board shall appoint an executive officer at a salary to be fixed and determined by the board with the approval of the Director of Finance.

This section shall become inoperative on July 1, 2011, and, as of January 1, 2012, is repealed, unless a later enacted statute, that becomes effective on or before January 1, 2012, deletes or extends the dates on which it becomes inoperative and is repealed.

6715. Roster

The board shall compile and maintain, or may have compiled and maintained on its behalf, a register of all licensees that contains information showing the name, address of record, type of branch license, license number, the date the license was issued, and the date the license will expire.

6716. Rules and Regulations; Meetings; Quorum

(a) The board may adopt rules and regulations consistent with law and necessary to govern its action. These rules and regulations shall be adopted in accordance with the provisions of the Administrative Procedure Act (Chapter 3.5 (commencing with Section 11340) of Part 1 of Division 3 of Title 2 of the Government Code.

(b) The board may adopt rules and regulations of professional conduct that are not inconsistent with state and federal laws. The rules and regulations may include definitions of incompetence and negligence. Every person who holds a license or certificate issued by the board pursuant to this chapter shall be governed by these rules and regulations.

(c) The board shall hold at least two regular meetings each year. Special meetings shall be held at those times that the board rules provide. A majority of the board constitutes a quorum.

6717. Authority to Define Scope of Practice

The board may, by regulation, define the scope of each branch of professional engineering other than civil, electrical, and mechanical engineering for which registration is provided under this chapter.

6718. Oaths and Testimony

Any member of the board may administer oaths and may take testimony and proofs concerning all matters within the board's jurisdiction.

6719. Board Seal

The board shall adopt and have an official seal which shall be affixed to all certificates of registration.

6720. Per Diem; Expenses

Each member of the board shall receive a per diem and expenses as provided in Section 103.

6726. Technical Advisory Committee; Functions

The board may establish one or more technical advisory committees to advise and assist the board with respect to the following:

(1) Application review and verification for any level of registration, licensure, authority, or title.

(2) Evaluation and investigation of potential violations of the act.

(3) Amendment, repeal, adoption, or revision of board rules, regulations, policies, and procedures.

6726.1. Membership of TAC

Each member of each technical advisory committee shall be appointed by the board and shall serve at the pleasure of the board. Each committee shall be composed of no more than five members.

6726.2. Qualification of Members

Each member of each technical advisory committee shall be an expert in the branch of engineering within the committee's jurisdiction and shall be licensed under this chapter.

6726.3. Per Diem; Expenses

All the members of each technical advisory committee shall serve without compensation but shall receive per diem and expenses as provided in Section 103.

6726.4. Immunity From Liability

Each member of each technical advisory committee shall be granted the same immunity as is granted to a public employee pursuant to Article 3 (commencing with Section 820) of Chapter 1 of Part 2 of Division 3.6 of Title 1 of the Government Code.

ARTICLE 2.3 PROFESSIONAL ENGINEERS REVIEW COMMITTEES

6728. Establishment; Duration

The board, when it deems necessary, may establish professional engineers review committees to hear all matters assigned by the board, including, but not limited to, any contested case which is assigned by the board. Each committee shall exist so long as the board deems that it is necessary.

6728.1. Members; Appointment; Qualifications

Each review committee shall consist of no fewer than three registered professional engineers appointed by the board. Each member of a committee shall have the same qualifications and shall be subject to the same rules and regulations as if he were a member of the board.

6728.2. Per Diem and Expenses

Each member of a committee shall receive a per diem and expenses as provided in Section 103 of this code.

6728.3. Hearings; Conduct; Presence of Hearing Officer

Except as otherwise provided in this article, all hearings which are conducted by a committee shall be conducted in accordance with the provisions of Chapter 4 (commencing with Section 11370), Chapter 4.5 (commencing with Section 11400), and Chapter 5 (commencing with Section 11500), Part 1, Division 3, Title 2 of the Government Code.

If a contested case is heard by a committee, the hearing officer who presided at the hearing shall be present during the committee's consideration of the case and, if requested, shall assist and advise the committee.

6728.4. Proposed Decisions

At the conclusion of any hearing which is conducted by a committee, the committee shall prepare a proposed decision, in such form that it may be adopted by the board as the decision in the case, and shall transmit it to the board. The proposed decision shall be subject to the same procedure as the proposed decision of a hearing officer under subdivisions (b) and (c) of Section 11517 of the Government Code.

6728.5. Authority for Rules and Regulations

The board may adopt, amend or repeal, in accordance with the provisions of Chapter 3.5 (commencing with Section 11340), Part 1, Division 3, Title 2 of the Government Code, such rules and regulations as are necessary to implement these sections.

6728.6. Immunity

Each member of a professional engineers review committee or other board-appointed committee and any board-appointed representative of the board shall be granted the same immunity as is granted to a public employee pursuant to Article 3 (commencing with Section 820) of Chapter 1 of Part 2 of Division 3.6 of Title 1 of the Government Code.

ARTICLE 3. APPLICATION OF CHAPTER

6730. Evidence of Qualifications; Registration

In order to safeguard life, health, property and public welfare, any person, either in a public or private capacity, except as in this chapter specifically excepted, who practices, or offers to practice, civil engineering, electrical engineering or mechanical engineering, in any of its branches in this state, including any person employed by the State of California, or any city, county, or city and county, who practices engineering, shall submit evidence that he is qualified to practice, and shall be licensed accordingly as a civil engineer, electrical engineer or mechanical engineer by the board.

6730.2. Requirement for Responsible Charge

It is the intent of the Legislature that the registration requirements that are imposed upon private sector professional engineers and engineering partnerships, firms, or corporations shall be imposed upon the state and any city, county, or city and county that shall adhere to those requirements. Therefore, for the purposes of Section 6730 and this chapter, at least one registered engineer shall be designated the person in responsible charge of professional engineering work for each branch of professional engineering practiced in any department or agency of the state, city, county, or city and county.

Any department or agency of the state or any city, county, or city and county which has an unregistered person in responsible charge of engineering work on January 1, 1985, shall be exempt from this requirement until that time as the person currently in responsible charge is replaced.

6731. Civil Engineering Defined

Civil engineering embraces the following studies or activities in connection with fixed works for irrigation, drainage, waterpower, water supply, flood control, inland waterways, harbors, municipal improvements, railroads, highways, tunnels, airports and airways, purification of water, sewerage, refuse disposal, foundations, grading, framed and homogeneous structures, buildings, or bridges:

(a) The economics of, the use and design of, materials of construction and the determination of their physical qualities.

(b) The supervision of the construction of engineering structures.

(c) The investigation of the laws, phenomena and forces of nature.

(d) Appraisals or valuations.

(e) The preparation or submission of designs, plans and specifications and engineering reports.

(f) Coordination of the work of professional, technical, or special consultants.

(g) Creation, preparation, or modification of electronic or computerized data in the performance of the activities described in subdivisions (a) through (f). Civil engineering also includes city and regional planning insofar as any of the above features are concerned therein. Civil engineers registered prior to January 1, 1982, shall be authorized to practice all land surveying as defined in Chapter 15 (commencing with Section 8700) of Division 3.

[NOTE: The last registration number issued to a civil engineer registered before January 1, 1982 was 33,965.]

6731.1. Civil Engineering; Additional Authority

Civil engineering also includes the practice or offer to practice, either in a public or private capacity, all of the following:

(a) Locates, relocates, establishes, reestablishes, or retraces the alignment or elevation for any of the fixed works embraced within the practice of civil engineering, as described in Section 6731.

(b) Determines the configuration or contour of the earth's surface or the position of fixed objects above, on, or below the surface of earth by applying the principles of trigonometry or photogrammetry.

(c) Creates, prepares, or modifies electronic or computerized data in the performance of the activities described in subdivisions (a) and (b).

(d) Renders a statement regarding the accuracy of maps or measured survey data pursuant to subdivisions (a), (b), and (c).

6731.2. Authority to Offer to Practice or Procure Land Surveying

Any registered civil engineer may offer to practice, procure, and offer to procure, land surveying work incidental to his or her civil engineering practice, even though he or she is not authorized to perform that work, provided all the land surveying work is performed by, or under the direction of, a licensed land surveyor or registered civil engineer authorized to practice land surveying. Further, any registered civil engineer may manage or conduct as manager, proprietor, or agent, a civil engineering practice which offers to practice, procure, and offers to procure, such incidental land surveying work.

6731.3 Construction Project Management Services

A registered civil engineer may also practice or offer to practice, either in a public or private capacity, construction project management services, including, but not limited to, construction project design review and evaluation, construction mobilization and supervision, bid evaluation, project scheduling, cost-benefit analysis, claims review and negotiation, and general management and administration of a construction project.

6731.4 Responsibility for Construction Management Services

If a registered civil engineer provides construction management services pursuant to Section 6731.3, Section 6703.1 shall not limit the responsibility of the engineer for the services actually provided.

6731.5. Electrical Engineering Defined

(a) Electrical engineering is that branch of professional engineering described in Section 6734.1 that embraces studies or activities relating to the generation, transmission, and utilization of electrical energy, including the design of electrical, electronic, and magnetic circuits, and the technical control of their operation and of the design of electrical gear. It is concerned with the research, organizational, and economic aspects of the above.

(b) The design of electronic and magnetic circuits is not exclusive to the practice of electrical engineering, as defined in subdivision (a).

6731.6. Mechanical Engineering Defined

Mechanical engineering is that branch of professional engineering described in Section 6734.2 that deals with engineering problems relating to generation, transmission, and utilization of energy in the thermal or mechanical form and also with engineering problems relating to the production of tools, machinery, and their products, and to heating, ventilation, refrigeration, and plumbing. It is concerned with the research, design, production, operational, organizational, and economic aspects of the above.

6732. Use of Seal, Stamp or Title by Unregistered Person

It is unlawful for anyone other than a professional engineer licensed under this chapter to stamp or seal any plans, specifications, plats, reports, or other documents with the seal or stamp of a professional engineer, or in any manner, use the title "professional engineer," "licensed engineer," "registered engineer," or "consulting engineer," or any of the following branch titles: "agricultural engineer," "chemical engineer," "civil engineer," "control system engineer," "electrical engineer," "fire protection engineer," "industrial engineer," "mechanical engineer," "metallurgical engineer," "nuclear engineer," "petroleum engineer," or "traffic engineer," or any combination of these words and phrases or abbreviations thereof unless licensed under this chapter.

6732.1. Exemption for Use of Title "Consulting Engineer"

Any person who has been granted permission to use the title "consulting engineer" pursuant to legislation enacted at the 1963, 1965, or 1968 Regular Session is exempt from the provisions of Section 6732 as it restricts the use of the title "consulting engineer", and such exemption shall apply so long as the applicant remains in practice and advises the board of any change of address within 30 days of such change. The board may adopt such rules under provisions of the Administrative Procedure Act as are necessary to implement this section.

The provisions of Articles 5 (commencing with Section 6775), 6 (commencing with Section 6785), and 7 (commencing with Section 6795) of this chapter shall apply to all persons who are granted permission to use the title "consulting engineer" pursuant to legislation enacted in 1963 and 1965 and the amendments to this section enacted at the 1968 Regular Session.

6732.2 Consulting Engineers; Practice of Photogrammetry

Any person who possesses a valid certificate to practice photogrammetry issued to him under the provisions of Chapter 15 (commencing with Section 8700) of this division may apply for, and be issued, a certificate of authority to use the title "consulting engineer," if all of the following requirements are satisfied:

(a) Application is made on a form provided by the board and is accompanied by the fees prescribed in Section 8805.

(b) Information submitted evidences to the satisfaction of the board that the applicant has had five years of independent control in furnishing consulting photogrammetric, geodetic, or topographic surveying services or consulting surveying services in connection with fixed works as defined in Section 6731.

Authority to use the title "consulting engineer" granted under this section does not affect authorizations made under the several provisions provided in Section 6732.1.

Authority to use the title "consulting engineer" granted under this section shall remain valid only while its holder's basic license is valid, and if it lapses it may be renewed only as provided in Article 7 (commencing with Section 6795).

The provisions of Article 5 (commencing with Section 6775), Article 6 (commencing with Section 6785), and Article 7 (commencing with Section 6795), of this chapter, shall apply to the certificates issued as provided in this section and to the persons so certificated.

6732.3. Continued Use of Branch Titles of Corrosion, Manufacturing, Quality, or Safety Engineering

(a) Any person who has received from the board a license in corrosion, manufacturing, quality, or safety engineering, and who holds a valid license under this chapter, may continue to use the branch title of the branch in which the professional engineer is legally licensed. A person holding a license in corrosion, manufacturing, quality, or safety engineering is subject to the license renewal provisions of this chapter.

(b) The professional engineer also may continue to use the title of "professional engineer," "licensed engineer," "registered engineer," or "consulting engineer."

6732.4. Examination for Registration as Corrosion, Manufacturing, Quality, or Safety Engineer

(a) Notwithstanding any other provision of law, any person who has applied for registration as a corrosion, quality, or safety engineer, and who has completed the written examination in one or more of these branch titles prior to January 1, 1999, shall be issued registration in the branch title for which the applicant was examined, provided that he or she has met all other qualifications for registration. The board shall not administer any examination for registration as a corrosion, quality, or safety engineer on or after January 1, 1999.

(b) Notwithstanding any other provision of law, any person who has applied for registration as a manufacturing engineer, and who has completed the written examination for this branch title prior to January 1, 2004, shall be issued a registration as a manufacturing engineer, provided that he or she has met all other qualifications for registration. The board shall not administer any examination for registration as a manufacturing engineer on or after January 1, 2004.

6732.5. Discontinuance of National Examination; Continued Use of Branch Titles

(a) Upon the discontinuance of a national examination for a branch specified in this chapter, the board shall not be required to administer an examination for a license in that branch or be required to issue licenses in that branch.

(b) Any person who has received from the board a license in a branch for which the nation examination is discontinued, and who holds a valid license under this chapter, may continue to use the branch title of the branch in which the professional engineer is legally licensed. A person holding a license in the affected branch of engineering is subject to the license renewal provisions of this chapter. The professional engineer may also continue to use the title of "professional engineer," "licensed engineer," "registered engineer," or "consulting engineer."

6733. Use of Stamp or Seal When Certificate Not in Force

It is unlawful for anyone to stamp or seal any plans, specifications, plats, reports, or other documents with the seal after the certificate of the registrant, named thereon, has expired or has been suspended or revoked, unless the certificate has been renewed or reissued.

6734. Practice of Civil Engineering

Any person practices civil engineering when he professes to be a civil engineer or is in responsible charge of civil engineering work.

6734.1. Practice of Electrical Engineering

Any person practices electrical engineering when he professes to be an electrical engineer or is in responsible charge of electrical engineering work.

6734.2. Practice of Mechanical Engineering

Any person practices mechanical engineering when he professes to be a mechanical engineer or is in responsible charge of mechanical engineering work.

6735. Preparation, Signing, and Sealing of Civil Engineering Documents

(a) All civil (including structural and geotechnical) engineering plans, calculations, specifications, and reports (hereinafter referred to as "documents") shall be prepared by, or under the responsible charge of, a registered civil engineer and shall include his or her name and license number. Interim documents shall include a notation as to the intended purpose of the document, such as "preliminary," "not for construction," "for plan check only," or "for review only." All civil engineering plans and specifications that are permitted or that are to be released for construction shall bear the signature and seal or stamp of the registrant, the date of signing and sealing or stamping, and the expiration date of the certificate or authority. All final civil engineering calculations and reports shall bear the signature and seal or stamp of the registrant, the date of signing and sealing or stamping, and the expiration date of the certificate or authority. If civil engineering plans are required to be signed and sealed or stamped and have multiple sheets, the signature, seal or stamp, date of signing and sealing or stamping, and expiration date of the certificate or authority shall appear on each sheet of the plans. If civil engineering specifications, calculations, and reports are required to be signed and sealed or stamped and have multiple pages, the signature, seal or stamp, date of signing and sealing or stamping, and expiration date of the certificate or authority shall appear at a minimum on the title sheet, cover sheet, or signature sheet.

(b) Notwithstanding subdivision (a), a registered civil engineer who signs civil engineering documents shall not be responsible for damage caused by subsequent changes to or uses of those documents, if the subsequent changes or uses, including changes or uses made by state or local governmental agencies, are not authorized or approved by the registered engineer who originally signed the documents, provided that the engineering service rendered by the civil engineer who signed the documents was not also a proximate cause of the damage.

6735.1. Construction Supervision; Legal Duty

The signing of civil engineering plans, specifications, reports, or documents which relate to the design of fixed works shall not impose a legal duty or responsibility upon the person signing the plans, specifications, reports, or documents to supervise the construction of engineering structures or the construction of the fixed works which are the subject of the plans, specifications, reports, or documents. However, nothing in this section shall preclude a civil engineer and a client from entering into a contractual agreement which includes a mutually acceptable arrangement for the provision of construction supervision services. Nothing contained in this subdivision shall modify the liability of a civil engineer who undertakes, contractually or otherwise, the provision of construction supervision services for rendering those services.

6735.3. Signing and Sealing of Electrical Engineering Documents

(a) All electrical engineering plans, specifications, calculations, and reports (hereinafter referred to as "documents") prepared by, or under the responsible charge of, a registered electrical engineer shall include his or her name and license number. Interim documents shall include a notation as to the intended purpose of the document, such as "preliminary," "not for construction," "for plan check only," or "for review only." All electrical engineering plans and specifications that are permitted or that are to be released for construction shall bear the signature and seal or stamp of the registrant, the date of signing and sealing or stamping, and the expiration date of the registration. All final electrical engineering calculations and reports shall bear the signature and seal or stamp of the registrant, the date of signing and sealing or stamping, and the expiration date of the registration. If electrical engineering plans are required to be signed and sealed or stamped and have multiple sheets, the signature, seal or stamp, date of signing and sealing or stamping, and expiration date of the registration shall appear on each sheet of the plans. If electrical engineering specifications, calculations, and reports are required to be signed and sealed or stamped and have multiple pages, the signature, seal or stamp, date of signing and sealing or stamping, and expiration date of the registration shall appear at a minimum on the title sheet, cover sheet, or signature sheet.

(b) Notwithstanding subdivision (a), a registered electrical engineer who signs electrical engineering documents shall not be responsible for damage caused by subsequent changes to or uses of those documents, if the subsequent changes or uses, including changes or uses made by state or local governmental agencies, are not authorized or approved by the registered engineer who originally signed the documents, provided that the engineering service rendered by the electrical engineer who signed the documents was not also a proximate cause of the damage.

6735.4. Signing and Sealing of Mechanical Engineering Documents

(a) All mechanical engineering plans, specifications, calculations, and reports (hereinafter referred to as "documents") prepared by, or under the responsible charge of, a registered mechanical engineer shall include his or her name and license number. Interim documents shall include a notation as to the intended purpose of the document, such as "preliminary," "not for construction," "for plan check only," or "for review only." All mechanical engineering plans and specifications that are permitted or that are to be released for construction shall bear the signature and seal or stamp of the registrant, the date of signing and sealing or stamping, and the expiration date of the registration. All final mechanical engineering calculations and reports shall bear the signature and seal or stamp of the registrant, the date of signing and sealing or stamping, and the expiration date of the registration. If mechanical engineering plans are required to be signed and sealed or stamped and have multiple sheets, the signature, seal or stamp, date of signing and sealing or stamping, and expiration date of the registration shall appear on each sheet of the plans. If mechanical engineering specifications, calculations, and reports are required to be signed and sealed or stamped and have multiple pages, the signature, seal or stamp, date of signing and sealing or stamping, and expiration date of the registration shall appear at a minimum on the title sheet, cover sheet, or signature sheet.

(b) Notwithstanding subdivision (a), a registered mechanical engineer who signs mechanical engineering documents shall not be responsible for damage caused by subsequent changes to or uses of those documents, if the subsequent changes or uses, including changes or uses made by state or local governmental agencies, are not authorized or approved by the registered engineer who originally signed the documents, provided that the engineering service rendered by the mechanical engineer who signed the documents was not also a proximate cause of the damage.

6735.5. Use of Word "Certify" or "Certification"

The use of the word "certify" or "certification" by a registered professional engineer in the practice of professional engineering or land surveying constitutes an expression of professional opinion regarding those facts or findings which are the subject of the certification, and does not constitute a warranty or guarantee, either expressed or implied.

6735.6. Provision of "As Built" Plans

If a registered civil engineer is required to provide as built, as constructed, or record plans for improvements or grading, which plans show changes during the construction process, the following shall apply:

(a) If the registered civil engineer provided construction phase services on the project that include supervision of the construction of engineering structures, the plans shall be based upon the field observations of the registered civil engineer and his or her agents, and information received from the project owner, project contractors, and public agencies.

(b) If the registered civil engineer did not provide construction phase services on the project that include supervision of the construction of engineering structures, the plans shall be based on information received from the project owner, project contractors, and public agencies, but need not be based upon a field verification or investigation of the improvements or grades, unless the registered civil engineer is engaged to provide such field verification services.

(c) The registered civil engineer shall not be required to include a certificate or statement on as built, as constructed, or record plans that is inconsistent with or varies from the provisions of this section.

6736. Title of Structural Engineer

No person shall use the title, "structural engineer," or any combination of these words or abbreviations thereof, unless he or she is a licensed civil engineer in this state and unless he or she has been found qualified as a structural engineer according to the rules and regulations established for structural engineers by the board.

6736.1. Soil Engineer, Soils Engineer, or Geotechnical Engineer

(a) No person shall use the title, "soil engineer," "soils engineer," or "geotechnical engineer," or any combination of these words or abbreviations thereof, unless he or she is a licensed civil engineer in this state and files an application to use the appropriate title with the board and the board determines the applicant is qualified to use the requested title.

(b) The board shall establish qualifications and standards to use the title "soil engineer," "soils engineer, or "geotechnical engineer." However, each applicant shall demonstrate a minimum of four years qualifying experience beyond that required for licensure as a civil engineer, and shall pass the examination specified by the board.

(c) For purposes of this section, "qualifying experience" means proof of responsible charge of soil engineering projects in at least 50 percent of the major areas of soil engineering, as determined by the board.

(d) Nothing contained in this chapter requires existing references to "soil engineering," "soils engineering," "geotechnical engineering," "soil engineer," "soils engineer," or "geotechnical engineer," in local agency ordinances, building codes, regulations, or policies, to mean that those activities or persons must be registered or authorized to use the relevant title or authority.

6737. Architectural Exemption

An architect, who holds a certificate to practice architecture in this State under the provisions of Chapter 3 of Division 3 of this code insofar as he practices architecture in its various branches, is exempt from registration under the provisions of this chapter.

6737.1. Structure Exemption

(a) This chapter does not prohibit any person from preparing plans, drawings, or specifications for any of the following:

(1) Single-family dwellings of woodframe construction not more than two stories and basement in height.

(2) Multiple dwellings containing no more than four dwelling units of woodframe construction not more than two stories and basement in height. However, this paragraph shall not be construed as allowing an unlicensed person to design multiple clusters of up to four dwelling units each to form apartment or condominium complexes where the total exceeds four units on any lawfully divided lot.

(3) Garages or other structures appurtenant to buildings described under subdivision (a), of wood-frame construction not more than two stories and basement in height.

(4) Agricultural and ranch buildings of woodframe construction, unless the building official having jurisdiction deems that an undue risk to the public health, safety or welfare is involved.

(b) If any portion of any structure exempted by this section deviates from substantial compliance with conventional framing requirements for woodframe construction found in the most recent edition of Title 24 of the California Code of Regulations or tables of limitation for woodframe construction, as defined by the applicable building code duly adopted by the local jurisdiction or the state, the building official having jurisdiction shall require the preparation of plans, drawings, specifications, or calculations for that portion by, or under the responsible charge of, a licensed engineer, or by, or under the responsible control of, an architect licensed pursuant to Chapter 3 (commencing with Section 5500). The documents for that portion shall bear the stamp and signature of the licensee who is responsible for their preparation.

6737.2. Supplementary Practice by Civil Engineer

Nothing in this chapter shall prohibit a civil engineer, registered under the provisions of this chapter, from practicing or offering to practice any engineering in connection with or supplementary to civil engineering studies or activities as defined in Section 6731.

6737.3. Exemption of Contractors

A contractor, licensed under Chapter 9 (commencing with Section 7000) of Division 3, is exempt from the provisions of this chapter relating to the practice of electrical or mechanical engineering so long as the services he or she holds himself or herself out as able to perform or does perform, which services are subject to the provisions of this chapter, are performed by, or under the responsible charge of a registered electrical or mechanical engineer insofar as the electrical or mechanical engineer practices the branch of engineering for which he or she is registered.

This section shall not prohibit a licensed contractor, while engaged in the business of contracting for the installation of electrical or mechanical systems or facilities, from designing those systems or facilities in accordance with applicable construction codes and standards for work to be performed and supervised by that contractor within the classification for which his or her license is issued, or from preparing electrical or mechanical shop or field drawings for work which he or she has contracted to perform. Nothing in this section is intended to imply that a licensed contractor may design work which is to be installed by another person.

6738. Engineering Business—Business Names

(a) This chapter does not prohibit one or more civil, electrical, or mechanical engineers from practicing or offering to practice within the scope of their license civil (including geotechnical and structural), electrical, or mechanical engineering as a sole proprietorship, partnership, firm, or corporation (hereinafter called business), if all of the following requirements are met:

(1) A civil, electrical, or mechanical engineer currently licensed in this state is an owner, partner, or officer in charge of the engineering practice of the business.

(2) All civil, electrical, or mechanical engineering services are performed by or under the responsible charge of a professional engineer licensed in the appropriate branch of professional engineering.

(3) If the business name of a California business contains the name of any person, then that person shall be licensed as a professional engineer, a licensed land surveyor, a licensed architect, or a geologist registered under the Geologist Act (Chapter 12.5 (commencing with Section 7800)). Any offer, promotion, or advertisement by the business that contains the name of any individual in the business, other than by use of the name of an individual in the business name, shall clearly and specifically designate the license or registration discipline of each individual named.

(b) An out-of-state business with a branch office in this state shall meet the requirements of subdivision (a) and shall have an owner, partner, or officer who is in charge of the engineering work in the branch in this state, who is licensed in this state, and who is physically present at the branch office in this state on a regular basis. However, the name of the business may contain the name of any person not licensed in this state if that person is appropriately registered or licensed in another state. Any offer, promotion, or advertisement which contains the name of any individual in the business, other than by use of the names of the individuals in the business name, shall clearly and specifically designate the license or registration discipline of each individual named.

(c) The business name of a California engineering business may be a fictitious name. However, if the fictitious name includes the name of any person, the requirements of paragraph (3) of subdivision (a) shall be met.

(d) A person not licensed under this chapter may also be a partner or an officer of a civil, electrical, or mechanical engineering business if the requirements of subdivision (a) are met. Nothing in this section shall be construed to permit a person who is not licensed under this chapter to be the sole owner of a civil, electrical, or mechanical engineering business, unless otherwise exempt under this chapter.

(e) This chapter does not prevent an individual or business engaged in any line of endeavor other than the practice of civil, electrical, or mechanical engineering from employing or contracting with a licensed civil, electrical, or mechanical engineer to perform the respective engineering services incidental to the conduct of business.

(f) This section shall not prevent the use of the name of any business engaged in rendering civil, electrical, or mechanical engineering services, including the use by any lawful successor or survivor, that lawfully was in existence on December 31, 1987. However, the business is subject to paragraphs (1) and (2) of subdivision (a).

(g) A business engaged in rendering civil, electrical, or mechanical engineering services may use in its name the name of a deceased or retired person provided all of the following conditions are satisfied:

(1) The person's name had been used in the name of the business, or a predecessor in interest of the business, prior to and after the death or retirement of the person.

(2) The person shall have been an owner, partner, or officer of the business, or an owner, partner, or officer of the predecessor in interest of the business.

(3) The person shall have been licensed as a professional engineer, or a land surveyor, or an architect, or a geologist, (A) by the appropriate licensing board if that person is operating a place of business or practice in this state, or (B) by the applicable state board if no place of business existed in this state.

(4) The person, if retired, has consented to the use of the name and does not permit the use of the name in the title of another professional engineering business in this state during the period of the consent. However, the retired person may use his or her name as the name of a new or purchased business if it is not identical in every respect to that person's name as used in the former business.

(5) The business shall be subject to the provisions of paragraphs (1) and (2) of subdivision (a).

(h) This section does not affect the provisions of Sections 6731.2 and 8726.1.

(i) A current organization record shall be filed with the board for all ~~business~~ businesses engaged in rendering civil, electrical, or mechanical engineering services.

{Amended by Stats.2007, Ch. 354}

6739. Exemption of Federal Officers and Employees

Officers and employees of the United States of America practicing solely as such officers or employees are exempt from registration under the provisions of this chapter.

6740. Exemption of Subordinates

A subordinate to a civil, electrical or mechanical engineer licensed under this chapter, or a subordinate to a civil, electrical or mechanical engineer exempted from licensure under this chapter, insofar as he acts solely in that capacity, is exempt from licensure under the provisions of this chapter. This exemption, however, does not permit any such subordinate to practice civil, electrical or mechanical engineering in his own right or to use the titles listed in Section 6732, 6736, and 6736.1.

6741. Exemption of Nonresidents

Any person, firm, partnership, or corporation is exempt from registration under the provisions of this chapter who meets all the following:

(a) Is a nonresident of the State of California.

(b) Is legally qualified in another state to practice as a civil, electrical, or mechanical engineer.

(c) Does not maintain a regular place of business in this state.

(d) Offers to but does not practice civil, electrical, or mechanical engineering in this state.

6742. Exemption for Real Estate Broker or Salesman

Any person, firm or corporation holding a license as real estate broker or real estate salesman, when making appraisals and valuations of real estate properties, while engaged in the business or acting in the capacity of a real estate broker or a real estate salesman, within the meaning of the California Real Estate Act is exempt from registration under the provisions of this chapter.

6743. Effect of Chapter on Surveyors

This chapter does not affect Chapter 15 of Division 3 of this code, relating to surveyors, except insofar as this chapter is expressly made applicable.

6744. Exemption for Land Owner

This chapter does not require registration for the purpose of practicing civil engineering, by an individual, a member of a firm or partnership, or by an officer of a corporation on or in connection with property owned or leased by the individual, firm, partnership, or corporation, unless the civil engineering work to be performed involves the public health or safety or the health and safety of employees of the individual, firm, partnership or corporation.

6745. Exemption for Building Alterations

This chapter does not prohibit any person, firm or corporation from furnishing, either alone or with subcontractors, labor and materials, with or without plans, drawings, specifications, instruments of service or other data covering such labor and materials:

(a) For store fronts, interior alterations or additions, fixtures, cabinet work, furniture or other appliances or equipment.

(b) For any work necessary to provide for their installation.

(c) For any alterations or additions to any building necessary to or attendant upon the installation of such store fronts, interior alterations or additions, fixtures, cabinet work, furniture, appliances or equipment; provided, such alterations do not affect the structural safety of the building.

6746. Exemption for Communications Companies Under the Public Utilities Commission

Plans, specifications, reports and documents relating to communication lines and equipment prepared by employees of communications companies which come under the jurisdiction of the Public Utilities Commission, and by employees of contractors while engaged in work on communication equipment for communications companies which come under the jurisdiction of the Public Utilities Commission, are not subject to the provisions of this chapter.

6746.1. Exemption for Employees of Communications Industry

The provisions of this act pertaining to licensure of professional engineers other than civil engineers, do not apply to employees in the communication industry, nor to the employees of contractors while engaged in work on communication equipment. However, those employees may not use any of the titles listed in Section 6732, 6736, and 6736.1, unless licensed.

6747. Exemption for Industries

(a) This chapter, except for those provisions that apply to civil engineers and civil engineering, shall not apply to the performance of engineering work by a manufacturing, mining, public utility, research and development, or other industrial corporation, or by employees of that corporation, provided that work is in connection with, or incidental to, the products, systems, or services of that corporation or its affiliates.

(b) For purposes of this section, "employees" also includes consultants, temporary employees, contract employees, and those persons hired pursuant to third-party contracts.

6748. Nuclear Power Plants

(a) Notwithstanding Section 6747, this chapter applies to engineering work approved by a person employed by a privately or publicly owned utility in the planning, designing, construction, operation, or maintenance of a nuclear powerplant which is owned or operated by the utility.

(b) Any engineering work subject to subdivision (a) shall be approved by a professional engineer registered in the applicable discipline of engineering specified in Section 6732.

6749. Written Contracts

(a) A professional engineer shall use a written contract when contracting to provide professional engineering services to a client pursuant to this chapter. The written contract shall be executed by the professional engineer and the client, or his or her representative, prior to the professional engineer commencing work, unless the client knowingly states in writing that work may be commenced before the contract is executed. The written contract shall include, but not be limited to, all of the following:

(1) A description of the services to be provided to the client by the professional engineer.

(2) A description of any basis of compensation applicable to the contract, and the method of payment agreed upon by the parties.

(3) The name, address, and license or certificate number of the professional engineer, and the name and address of the client.

(4) A description of the procedure that the professional engineer and the client will use to accommodate additional services.

(5) A description of the procedure to be used by any party to terminate the contract.

(b) This section shall not apply to any of the following:

(1) Professional engineering services rendered by a professional engineer for which the client will not pay compensation.

(2) A professional engineer who has a current or prior contractual relationship with the client to provide engineering services, and that client has paid the professional engineer all of the fees that are due under the contract.

(3) If the client knowingly states in writing after full disclosure of this section that a contract which complies with the requirements of this section is not required.

(4) Professional engineering services rendered by a professional engineer to any of the following:

(A) A professional engineer licensed or registered under this chapter.

(B) A land surveyor licensed under Chapter 15 (commencing with Section 8700).

(C) An architect licensed under Chapter 3 (commencing with Section 5500).

(D) A contractor licensed under Chapter 9 (commencing with Section 7000).

(E) A geologist or a geophysicist licensed under Chapter 12.5 (commencing with Section 7800).

(F) A manufacturing, mining, public utility, research and development, or other industrial corporation, if the services are provided in connection with or incidental to the products, systems, or services of that corporation or its affiliates.

(G) A public agency.

(c) "Written contract" as used in this section includes a contract that is in electronic form.

ARTICLE 4. REGISTRATION

6750. Application; Fee

(a) An application for licensure as a professional engineer or certification as an engineerin-training shall be made to the board on the prescribed form, with all statements made therein under oath, and shall be accompanied by the application fee prescribed by this chapter. An application for licensure as a professional engineer shall specify, additionally, the branch of engineering in which the applicant desires licensure.

(b) The board may authorize an organization specified by the board pursuant to Section 6754 to receive directly from applicants payment of the examination fees charged by that organization as payment for examination materials and services.

6751. Qualifications

(a) The applicant for certification as an engineer-in-training shall comply with all of the following:

(1) Not have committed acts or crimes constituting grounds for denial of registration under Section 480.

(2) Successfully pass the first division of the examination. The applicant shall be eligible to sit for the first division of the examination after satisfactory completion of three years or more of college or university education in a board-approved engineering curriculum or after completion of three years or more of board-approved experience.

The board need not verify the applicant's eligibility other than to require the applicant to sign a statement of eligibility on the application form.

(b) The applicant for registration as a professional engineer shall comply with all of the following:

(1) Not have committed acts or crimes constituting grounds for denial of registration under Section 480.

(2) Furnish evidence of six years or more of qualifying experience in engineering work satisfactory to the board evidencing that the applicant is competent to practice the character of engineering in the branch for which he or she is applying for registration, and successfully pass the second division of the examination.

(3) The applicant for the second division of the examination shall successfully pass the first division examination or shall be exempt therefrom.

6751.2. Foreign Applicants

The board may consider the professional experience and education acquired by applicants outside the United States which in the opinion of the board is equivalent to the minimum requirements of the board established by regulation for professional experience and education in this state.

6751.5. Rules for Approval of Engineering School Curricula

The board shall by rule establish the criteria to be used for approving curricula of schools of engineering.

6752. Civil Engineer Experience

An applicant for registration as a civil engineer must have gained his experience under the direction of a civil engineer legally qualified to practice.

6753. Equivalents for Experience; Education; Teaching

With respect to applicants for licensure as professional engineers, the board:

(a) Shall give credit as qualifying experience of four years, for graduation with an engineering degree from a college or university the curriculum of which has been approved by the board.

(b) May at its discretion give credit as qualifying experience up to a maximum of two years, for graduation with an engineering degree from a nonapproved engineering curriculum or graduation with an engineering technology degree in an approved engineering technology curriculum.

(c) May at its discretion give credit as qualifying experience of up to one-half year, for each year of successfully completed postsecondary study in an engineering curriculum up to a maximum of four years credit. A year of study shall be at least 32 semester units or 48 quarter units.

(d) May at its discretion give credit as qualifying experience not in excess of five years, for a postgraduate degree in a school of engineering with a board-approved undergraduate or postgraduate curriculum.

(e) May at its discretion give credit as qualifying experience for engineering teaching, not in excess of one year, if of a character satisfactory to the board.

The sum of qualifying experience credit for subdivision (a) to (e), inclusive, shall not exceed five years.

6753.5. Experience in Armed Forces

All applicants shall be given equal credit for engineering experience in the armed forces of United States as with any other comparable engineering experience.

6754. Examinations in General

Examination for licensure shall be held at such times and places as the board shall determine.

The second division of the examination for all branches specified in Section 6732 shall be administered at least once each year.

Work of the board relating to examination and licensure may be divided into committees as the board shall direct. The scope of examinations and the methods of procedure may be prescribed by board rule.

The board may make arrangements with a public or private organization to conduct the examination. The board may contract with a public or private organization for materials or services related to the examination.

6755. Examination Requirements

(a) Examination duration and composition shall be designed to conform to the following general principle: The first division of the examination shall test the applicant's knowledge of appropriate fundamental engineering subjects, including mathematics and the basic sciences; the second division of the examination shall test the applicant's ability to apply his or her knowledge and experience and to assume responsible charge in the professional practice of the branch of engineering in which the applicant is being examined.

(b) The applicant for the second division of the examination shall have successfully passed the first division examination or shall be exempt therefrom.

(c) The board may by rule provide for a waiver of the first division of the examination for applicants whose education and experience qualifications substantially exceed the requirements of Section 6751.

(d) The board may by rule provide for a waiver of the second division of the examination for persons eminently qualified for registration in this state by virtue of their standing in the engineering community, their years of experience, and those other qualifications as the board deems appropriate.

6755.1. Second Division Exam Requirements

(a) The second division of the examination for registration as a professional engineer shall include questions to test the applicant's knowledge of state laws and the board's rules and regulations regulating the practice of professional engineering. The board shall administer the test on state laws and board rules regulating the practice of engineering in this state as a separate part of the second division of the examination for registration as a professional engineer.

(b) On and after April 1, 1988, the second division of the examination for registration as a civil engineer shall also include questions to test the applicant's knowledge of seismic principles and engineering surveying principles as defined in Section 6731.1. No registration for a civil engineer shall be issued by the board on or after January 1, 1988, to any applicant unless he or she has successfully completed questions to test his or her knowledge of seismic principles and engineering surveying principles.

The board shall administer the questions to test the applicant's knowledge of seismic principles and engineering surveying principles as a separate part of the second division of the examination for registration as a civil engineer.

It is the intent of the Legislature that this section confirm the authority of the board to issue registrations prior to April 1, 1988, to applicants based on examinations not testing the applicant's knowledge of seismic principles and engineering surveying principles as defined in Section 6731.1.

6756. Certification as Engineer-in-Training

(a) An applicant for certification as an engineer-in-training shall, upon making a passing grade in that division of the examination prescribed in Section 6755, relating to fundamental engineering subjects, be issued a certificate as an engineer-in-training. A renewal or other fee, other than the application fee, may not be charged for this certification. The certificate shall become invalid when the holder has qualified as a professional engineer as provided in Section 6762.

(b) An engineer-in-training certificate does not authorize the holder thereof to practice or offer to practice civil, electrical or mechanical engineering work, in his or her own right, or to use the titles specified in Sections 6732, 6736, and 6736.1.

(c) It is unlawful for anyone other than the holder of a valid engineer-in-training certificate issued under this chapter to use the title of "engineer-in-training" or any abbreviation of that title.

6757. Separate Branches of Engineering

Applicants who profess to be qualified in more than one branch of engineering shall be required to file an application for each branch in which they wish to be registered.

6758. Applications, Reexamination

An applicant failing in an examination may be examined again upon filing a new application and the payment of the application fee fixed by this chapter.

6759. Comity Applicants

The board, upon application therefor, on its prescribed form, and the payment of the application fee fixed by this chapter, may issue a certificate of registration as a professional engineer, without written examination, to any person holding a certificate of registration issued to him or her by any state or country when the applicant's qualifications meet the requirements of this chapter, and rules established by the board. The board shall not require a comity applicant to meet any requirement not required of California applicants. For purposes of this section, equivalent second division examinations shall be eight-hour written examinations prepared by or administered by a state or territory either by single or combined branch at the level generally administered by the board to persons who passed or were exempted from the first division examination. Applicants who have passed an equivalent second division combined branch or a single branch examination in a branch not recognized for registration in California shall be registered in the branch in which their experience and education indicate the closest relationship.

6760. Temporary Authorization for Practice

A temporary authorization to practice engineering in a branch defined by this chapter may be granted for a specific project, upon application and payment of the fee prescribed in Section 6799, for a period not to exceed 180 consecutive days, if the applicant complies with all of the following:

(a) The applicant maintains no place of business in this state.

(b) The applicant is legally qualified to practice the branch of engineering in which he or she is seeking the temporary authorization in the state or country where he or she maintains a place of business.

(c) (1) The applicant, if applying for a temporary authorization to practice civil engineering, demonstrates by means of an individual appearance before the board satisfactory evidence of his or her knowledge of the application of seismic forces in the design of structures or adequate knowledge in any of the other phases of civil engineering as related to the specific project for which the temporary authorization is requested.

(2) The applicant, if applying for a temporary authorization to practice in a branch defined by this chapter other than civil engineering, demonstrates by means of an individual appearance before the board, satisfactory evidence of his or her knowledge in the branch of professional engineering in which the applicant proposes to practice under the temporary authorization as related to the specific project for which the temporary authorization is requested.

(d) The applicant takes and passes the examination in the state laws and board rules prescribed in Section 6755.1.

(e) The applicant notifies the board in writing of his or her intention to practice, stating the approximate date he or she intends to commence the specific project and the approximate duration of the specific project, which shall not exceed 180 consecutive days from the commencement date of the specific project.

Upon completion of the requirements, the executive officer, on the direction of the board, shall issue a temporary authorization to the applicant.

6761. Majority Vote on Qualifications

In determining the qualifications of an applicant for registration, a majority vote of the board is required.

6762. Certification as Professional Engineer

Any applicant who has passed the second division examination and has otherwise qualified hereunder as a professional engineer, shall have a certificate of registration issued to him or her as a professional engineer in the particular branch for which he or she is found qualified.

6762.5. Retired License

(a) The board shall issue, upon application and payment of the fee established by Section 6799, a retired license (registration), to an engineer who has been licensed by the board for a minimum of 5 years within

California and a minimum of 20 years within the United States or territory of the United States, and who holds a license that is not suspended, revoked, or otherwise disciplined, or subject to pending discipline under this chapter.

(b) The holder of a retired license issued pursuant to this section shall not engage in any activity for which an active engineer's license is required. An engineer holding a retired license shall be permitted to use the titles "retired professional engineer," "professional engineer, retired," or either of those titles with the licensee's branch designation inserted for the word "professional" for example, "retired civil engineer" or "civil engineer, retired."

(c) The holder of a retired license shall not be required to renew that license.

(d) In order for the holder of a retired license issued pursuant to this section to restore his or her license to active status, he or she shall pass the second division examination that is required for initial licensure with the board.

6763. Structural, Soil, Soils, Geotechnical Authority

Application for authority to use the title "structural engineer," "soil engineer," "soils engineer," or "geotechnical engineer" shall be made to the board on forms prescribed by it and shall be accompanied by the application fee fixed by this chapter.

An applicant for authority to use the title "structural engineer," "soil engineer," "soils engineer," or "geotechnical engineer" who has passed the examination prescribed by the board, or an applicant for authority to use the title "soil engineer," "soils engineer," or "geotechnical engineer" whose application is submitted prior to July 1, 1986, and who has otherwise demonstrated that he or she is qualified, shall have a certificate of authority issued to him or her.

For purposes of this chapter, an authority to use the title "structural engineer," "soil engineer," "soils engineer," or "geotechnical engineer" is an identification of competence and specialization in a subspecialty of civil engineering and necessitates education or experience in addition to that required for registration as a civil engineer.

6763.1. Structural Engineer Examination Requirements

An applicant to use the title "structural engineer" shall have successfully passed a written examination that incorporates a national examination for structural engineering by a nationally recognized entity approved by the board, if available, and a supplemental California specific examination. The California specific examination shall test the applicant's knowledge of state laws, rules, and regulations, and of seismicity and structural engineering unique to practice in this state. The board shall use the national examination on or before December 31, 2004.

6763.5. Refund to Unqualified Applicants

If an applicant for registration as a professional engineer or certification as an engineerin-training, or for authorization to use the title "structural engineer" or "soil engineer," is found by the board to lack the qualifications required for admission to the examination for such registration, certification, or authorization, the board may, in accordance with the provisions of Section 158 of this code, refund to him or her one-half of the amount of his or her application fee.

6764. Seal or Stamp

Each professional engineer registered under this chapter shall, upon registration, obtain a seal or stamp of a design authorized by the board bearing the registrant's name, number of his or her certificate or authority, the legend "professional engineer" and the designation of the particular branch or authority in which he or she is registered, and may bear the expiration date of the certificate or authority.

6765. Duplicate Certificate

A duplicate certificate of registration to replace one lost, destroyed, or mutilated may be issued subject to the rules and regulations of the board. The duplicate certificate fee fixed by this chapter shall be charged.

6766. Certificate as Evidence

An unsuspended, unrevoked and unexpired certificate and endorsement of registry made under this chapter, is presumptive evidence in all courts and places that the person named therein is legally registered.

ARTICLE 4.5. REPORTING REQUIREMENTS

6770. Licensees Required to Report Criminal Convictions, Civil Action Judgments, Settlements, Arbitration Awards, and Administrative Actions

(a) A licensee shall report to the board in writing the occurrence of any of the following events that occurred on or after ~~July 1, 2006~~ January 1, 2008, within 90 days of the date the licensee has knowledge of the event:

(1) The conviction of the licensee of any felony.

(2) The conviction of the licensee of any other crime that is substantially related to the qualifications, functions, and duties of a licensed professional engineer.

(3) Any civil action judgment, settlement, arbitration award, or administrative action resulting in a judgment, settlement, or arbitration award against the licensee in any action alleging fraud, deceit, misrepresentation, breach or violation of contract, negligence, incompetence, or recklessness by the licensee in the practice of professional engineering if the amount or value of the judgment, settlement, or arbitration award is fifty thousand dollars ($50,000) or greater.

(b) The report required by subdivision (a) shall be signed by the licensee and set forth the facts that constitute the reportable event. If the reportable event involves the action of an administrative agency or court, the report shall set forth the title of the matter, court or agency name, docket number, and the date the reportable event occurred.

(c) A licensee shall promptly respond to oral or written inquiries from the board concerning the reportable events, including inquiries made by the board in conjunction with license renewal.

(d) Nothing in this section shall impose a duty upon any licensee to report to the board the occurrence of any of the events set forth in subdivision (a) either by or against any other licensee.

(e) Failure of a licensee to report to the board in the time and manner required by this section shall be grounds for disciplinary action.

(f) For the purposes of this section, a conviction includes the initial plea, verdict, or finding of guilt; a plea of no contest; or pronouncement of sentence by a trial court even though the conviction may not be final or sentence actually imposed until all appeals are exhausted.

6770.1. Courts Required to Report of Convictions of Licensees

Within 30 days of entry of a conviction described in paragraphs (1) and (2) of subdivision (a) of Section 6770 or a judgment described in paragraph (3) of subdivision (a) of Section 6770 by a court of this state that has been notified that the defendant is a licensee of the board, the court that rendered the conviction or judgment shall report that fact to the board and provide the board with a copy of the conviction or judgment and any orders or opinions of the court accompanying or ordering the conviction or judgment.

6770.2. Professional Liability Insurers Required to Report of Civil Action Judgments, Settlements, or Arbitration Awards Against Licensees

(a) Within 30 days of payment of all or any portion of any civil action judgment, settlement, or arbitration award described in Section 6770 against a licensee of the board in which the amount or value of the judgment, settlement, or arbitration award is fifty thousand dollars ($50,000) or greater, any insurer providing professional liability insurance to that licensee shall report to the board the name of the licensee; the amount or value of the judgment, settlement, or arbitration award; the amount paid by the insurer; and the identity of the payee.

(b) Within 30 days of payment of all or any portion of any civil action judgment, settlement, or arbitration award described in Section 6770 against a licensee of the board in which the amount or value of the judgment, settlement, or arbitration award is fifty thousand dollars ($50,000) or greater, any state or local government agency that self insures that licensee shall report to the board the name of the licensee; the amount or value of the judgment, settlement, or arbitration award; the amount paid; and the identity of the payee.

6770.3. Applicability of Sections 6770, 6770.1, and 6770.2

The requirements of Sections 6770, 6770.1, and 6770.2 shall apply if a party to the civil action, settlement, or arbitration award is or was a sole proprietorship, partnership, firm, corporation, or state or local government agency in which the licensee is or was an owner, partner, member, officer, or employee and is or was the licensee in responsible charge of that portion of the project that was the subject of the civil judgment, settlement, or arbitration award.

6770.4. Effect of Article on Confidential Agreements

(a) Notwithstanding any other provision of law, a licensee shall not be considered to have violated a confidential settlement agreement or other confidential agreement by providing a report to the board as required by this article.

6770.5. Authority to Adopt Regulations

The board may adopt regulations to further define the reporting requirements of Sections 6770, 6770.1, and 6770.2.

6770.6. Operation of Article

This article shall become operative on ~~July 1, 2006~~ January 1, 2008, only if an appropriation is made from the Professional Engineer's and Land Surveyor's Fund for the ~~200607~~ 2007-08 fiscal year in the annual Budget Act to fund the activities of this article, and sufficient hiring authority is granted to the board pursuant to a budget change proposal to provide sufficient staffing to implement this article.

{Amended by Stats.2007, Ch. 354}

ARTICLE 5. DISCIPLINARY PROCEEDINGS

6775. Complaints Against Professional Engineers

The Board may receive and investigate complaints against registered professional engineers, and make findings thereon.

By a majority vote, the board may reprove, suspend for a period not to exceed two years, or revoke the certificate of any professional engineer registered under this chapter:

(a) Who has been convicted of a crime substantially related to the qualifications, functions and duties of a registered professional engineer, in which case the certified record of conviction shall be conclusive evidence thereof.

(b) Who has been found guilty by the board of any deceit, misrepresentation, or fraud, in his or her practice.

(c) Who has been found guilty by the board of negligence or incompetence in his or her practice.

(d) Who has been found guilty by the board of any breach or violation of a contract to provide professional engineering services.

(e) Who has been found guilty of any fraud or deceit in obtaining his or her certificate.

(f) Who aids or abets any person in the violation of any provision of this chapter.

(g) Who in the course of the practice of professional engineering has been found guilty by the board of having violated a rule or regulation of unprofessional conduct adopted by the board.

(h) Who violates any provision of this chapter.

6775.1. Complaints Against Engineers-in-Training

The board may receive and investigate complaints against engineers-in-training, and make findings thereon.

By a majority vote, the board may revoke the certificate of any engineer-in-training:

(a) Who has been convicted of a crime as defined in subdivision (a) of Section 480.

(b) Who has committed any act that would be grounds for denial of licensure pursuant to Section 480 or 496.

(c) Who has been found guilty of any fraud, deceit, or misrepresentation in obtaining his or her engineer-in-training certificate or certificate of registration, certification, or authority as a professional engineer.

(c) (d) Who aids or abets any person in the violation of any provision of this chapter.

(d) (e) Who violates Section 119 with respect to an engineer-in-training certificate. or

(f) Who commits any act described in Section 6787.

(e) (g) Who violates any provision of this chapter.

{Amended by Stats.2007, Ch. 354}

6776. Conduct of Proceedings

The proceedings under this article shall be conducted in accordance with Chapter 4 (commencing with Section 11370), Chapter 4.5 (commencing with Section 11400), and Chapter 5 (commencing with Section 11500) of Part 1 of Division 3 of Title 2 of the Government Code, and the board shall have all the powers granted therein.

6777. Reissue of Certificate

The board may reissue a certificate of registration, certification, or authority, to any person whose certificate has been revoked if a majority of the members of the board vote in favor of such reissuance for reasons the board deems sufficient.

6779. Revocation Upon Conviction

A plea or verdict of guilty or a conviction following a plea of nolo contendere made to a charge substantially related to the qualifications, functions and duties of a registered professional engineer is deemed to be a conviction within the meaning of this article. The board may order the certificate suspended or revoked, or may decline to issue a certificate, when the time for appeal has elapsed, or the judgment of conviction has been affirmed on appeal or when an order granting probation is made suspending the imposition of sentence, irrespective of a subsequent order under the provisions of Section 1203.4 of the Penal Code allowing such person to withdraw his plea of guilty and to enter a plea of not guilty, or setting aside the verdict of guilty, or dismissing the accusation, information or indictment.

6780. Petitions for Reinstatement or Modification of Penalty

(a) A petitioner may petition the board for reinstatement or modification of penalty, including reduction, modification, or termination of probation, after the following minimum periods have elapsed from the effective date of the decision ordering the disciplinary action, or if the order of the board or any portion of it is stayed by a court of law, from the date the disciplinary action is actually implemented in its entirety:

(1) Except as otherwise provided in this section, at least three years for reinstatement of a certificate that was revoked or surrendered. However, the board may, in its sole discretion, specify in its order of revocation or surrender a lesser period of time that shall be at minimum one year.

(2) At least two years for early termination of a probation period of three years or more.

(3) At least one year for early termination of a probation period of less than three years.

(4) At least one year for reduction or modification of a condition of probation.

(b) The board shall notify the Attorney General of the filing of the petition. The petitioner and the Attorney General shall be given timely notice by letter of the time and place of the hearing on the petition, and the petitioner and the Attorney General shall be given the opportunity to present both oral and documentary evidence and argument to the board. The petitioner shall at all times have the burden of proof to establish by clear and convincing evidence that he or she is entitled to the relief sought in the petition.

(c) The board itself or an administrative law judge, if one is designated by the board, shall hear the petition and shall prepare a written decision setting forth the reasons supporting the decision.

(d) The board may grant or deny the petition or may impose any terms and conditions that it reasonably deems appropriate as a condition of reinstatement or reduction or modification of the penalty.

(e) No petition shall be considered while the petitioner is under sentence for any criminal offense, including any period during which the petitioner is on court-imposed probation or parole. No petition shall be considered while there is an accusation or petition to revoke probation pending against the petitioner.

(f) The board may, in its discretion, deny without hearing or argument any petition that is filed pursuant to this section within a period of two years from the effective date of a prior decision following a hearing under this section.

(g) Judicial review of the board's decision following a hearing under this section may be sought by way of a petition for writ of administrative mandamus pursuant to Section 1094.5 of the Code of Civil Procedure. The party seeking to overturn the board's decision shall have the burden of proof in any mandamus proceeding. In the mandamus proceeding, if it is alleged that there has been an abuse of discretion because the board's findings are not supported by the evidence, abuse of discretion is established if the court determines that the findings are not supported by substantial evidence in light of the whole record.

(h) The following definitions apply for purposes of this section:

(1) "Certificate" includes certificate of registration or license as a professional engineer; certificates of authority to use the titles "structural engineer," "geotechnical engineer," "soil engineer," "soils engineer," or "consulting engineer;" and certification as an engineer-in-training.

(2) "Petitioner" means a professional engineer or an engineer-in-training whose certificate has been revoked, suspended, or surrendered or placed on probation.

ARTICLE 6. OFFENSES AGAINST THE CHAPTER

6785. Investigations

The board shall have the power, duty, and authority to investigate violations of the provisions of this chapter.

6786. Duty to Prosecute

It is the duty of the respective members of the legal system to prosecute all persons charged with the violation of any of the provisions of this chapter.

It is the duty of the executive officer of the board, under the direction of the board, to aid these officers in the enforcement of this chapter.

6787. Acts Constituting Misdemeanor

Every person is guilty of a misdemeanor:

(a) Who, unless he or she is exempt from licensure under this chapter, practices or offers to practice civil, electrical, or mechanical engineering in this state according to the provisions of this chapter without legal authorization.

(b) Who presents or attempts to file as his or her own the certificate of licensure of a licensed professional engineer unless he or she is the person named on the certificate of licensure.

(c) Who gives false evidence of any kind to the board, or to any member thereof, in obtaining a certificate of licensure.

(d) Who impersonates or uses the seal of a licensed professional engineer.

(e) Who uses an expired, suspended, or revoked certificate issued by the board.

(f) Who represents himself or herself as, or uses the title of, a licensed or registered civil, electrical, or mechanical engineer, or any other title whereby that person could be considered as practicing or offering to practice civil, electrical, or mechanical engineering in any of its branches, unless he or she is correspondingly qualified by licensure as a civil, electrical, or mechanical engineer under this chapter.

(g) Who, unless appropriately registered, manages, or conducts as manager, proprietor, or agent, any place of business from which civil, electrical, or mechanical engineering work is solicited, performed, or practiced, except as authorized pursuant to subdivision (d) of Section 6738 and Section 8726.1.

(h) Who uses the title, or any combination of that title, of "professional engineer," "licensed engineer," "registered engineer," or the branch titles specified in Section 6732, or the authority titles specified in Sections 6736 and 6736.1, or "engineer-in-training," or who makes use of any abbreviation of such title that might lead to the belief that he or she is a licensed engineer, is authorized to use the titles specified in Section 6736 or 6736.1, or holds a certificate as an engineer-in-training, without being licensed, authorized, or certified as required by this chapter.

(i) Who uses the title "consulting engineer" without being licensed as required by this chapter or without being authorized to use that title pursuant to legislation enacted at the 1963, 1965 or 1968 Regular Session.

(j) Who violates any provision of this chapter.

6788. Repair Fraud

Any person who violates any provision of subdivisions (a) to (i), inclusive, of Section 6787 in connection with the offer or performance of engineering services for the repair of damage to a residential or nonresidential structure caused by a disaster for which a state of emergency is proclaimed by the Governor pursuant to Section 8625 of the Government Code, or for which an emergency or major disaster is declared by the President of the United States, shall be punished by a fine up to ten thousand dollars ($10,000), or by imprisonment in the state prison for 16 months, or for two or three years, or by both the fine and imprisonment, or by a fine up to one thousand dollars ($1,000), or by imprisonment in the county jail not exceeding one year, or by both the fine and imprisonment.

ARTICLE 7. REVENUE

6795. Certificate Renewal Time Periods

Certificates of registration as a professional engineer, and certificates of authority, shall be valid for a period of two years from the assigned date of renewal. Biennial renewals shall be staggered on a monthly basis. To renew an unexpired certificate, the certificate holder shall, on or before the date of expiration indicated on the renewal receipt, apply for renewal on a form prescribed by the board, and pay the renewal fee prescribed by this chapter.

6795.1. Notice of Pending Expiration; Contents; Second Notice

Within 60 to 90 days prior to the expiration of a certificate of registration or certificate of authority, the board shall mail to the registrant or authority holder a notice of the pending expiration. That notice shall include application forms for renewal. If there is no response by the expiration date, the board shall provide a second notice to the registrant's or authority holder's address.

6796. Renewal of Expired Certificate

Except as otherwise provided in this article, certificates of registration as a professional engineer, and certificates of authority may be renewed at any time within three years after expiration on filing of application for renewal on a form prescribed by the board and payment of all accrued and unpaid renewal fees. If the certificate is renewed more than 60 days after its expiration, the certificate holder, as a condition precedent to renewal, shall also pay the delinquency fee prescribed by this chapter. Renewal under this section shall be effective on the date on which the application is filed, on the date on which the renewal fee is paid, or on the date on which the delinquency fee, if any, is paid, whichever last occurs.

The expiration date of a certificate renewed pursuant to this section shall be determined pursuant to Section 6795.

6796.1. Expiration of Suspended Certificate; Renewal

A suspended certificate is subject to expiration and shall be renewed as provided in this article, but such renewal does not entitle the holder of the certificate, while it remains suspended and until it is reinstated, to engage in the activity to which the certificate relates, or in any other activity or conduct in violation of the order or judgment by which it was suspended.

6796.2. Expiration of Revoked Certificate

A revoked certificate is subject to expiration as provided in this article, but it may not be renewed. If it is reinstated after its expiration, the holder of the certificate, as a condition precedent to its reinstatement, shall pay a reinstatement fee in an amount equal to the renewal fee in effect on the last regular renewal date before the date on which it is reinstated, plus the delinquency fee, if any, accrued at the time of its revocation.

6796.3. Delinquent Registration, Over Three Years

Certificates of registration as a professional engineer, and certificates of authority to use the title "structural engineer," "soil engineer," or "consulting engineer" that are not renewed within three years after expiration may not be renewed, restored, reinstated, or reissued unless all of the following apply:

(a) The registrant or certificate holder has not committed any acts or crimes constituting grounds for denial of registration or of a certificate under Section 480.

(b) The registrant or certificate holder takes and passes the examination that would be required of him or her if he or she were then applying for the certificate for the first time, or otherwise establishes to the satisfaction of the board that, with due regard for the public interest, he or she is qualified to practice the branch of engineering in which he or she seeks renewal or reinstatement.

(c) The registrant or certificate holder pays all of the fees that would be required of him or her if he or she were then applying for the certificate for the first time. If the registrant or certificate holder has been practicing in this state with an expired or delinquent license and receives a waiver from taking the examination as specified in subdivision (b) then he or she shall pay all accrued and unpaid renewal fees.

The board may, by regulation, provide for the waiver or refund of all or any part of the application fee in those cases in which a certificate is issued without an examination pursuant to this section.

6796.5. Effect of Renewal of Expired or Delinquent Certificate

Once an expired or delinquent certificate of registration or certificate of authority is renewed, restored, reinstated, or reissued pursuant to Section 6796 or 6796.3, all of the following shall apply:

(a) The board shall continue to have full jurisdiction and authority over the registrant or certificate holder as if the registration or authority had not expired or become delinquent.

(b) The work performed by the registrant or certificate holder during a period of expiration or delinquency shall be deemed lawful and validly performed as to persons or entities other than the registrant or authority holder.

(c) The renewal, restoration, reinstatement, or reissuance of a registration or certificate of authority shall not affect liability issues regarding work performed during a period of expiration or delinquency, nor does the fact of performance during a period of expiration on delinquency affect liability issues.

6797. Accounting and Deposit of Funds

The department shall receive and account for all money derived from the operation of this chapter and, at the end of each month, shall report such money to the State Controller and shall pay it to the State Treasurer, who shall keep the money in a separate fund known as the Professional Engineer's and Land Surveyor's Fund.

For accounting and record keeping purposes, the Professional Engineer's and Land Surveyor's Fund shall be deemed to be a single special fund, and shall be available for expenditure only for the purposes as are now or may hereafter be provided by law.

The fees and civil penalties received under this chapter shall be deposited in the Professional Engineer's and Land Surveyor's Fund. All moneys in the fund are hereby appropriated for the purposes of this chapter.

6798. Refunds

The board may make refunds of all fees in accordance with Section 158 of this code.

6799. Fee Schedule

The amount of the fees prescribed by this chapter shall be fixed by the board in accordance with the following schedule:

(a) The fee for filing each application for ~~registration~~ licensure as a professional engineer and each application for authority level designation at not more than four hundred dollars ($400), and for each application for certification as an engineer-in-training at not more than one hundred dollars ($100).

(b) The temporary ~~registration~~ authorization fee for a professional engineer at not more than 25 percent of the application fee in effect on the date of application.

(c) The renewal fee for each branch of professional engineering in which ~~registration~~ licensure is held, and the renewal fee for each authority level designation held, at no more than the professional engineer application fee currently in effect.

(d) The fee for a retired license at not more than 50 percent of the professional engineer application fee in effect on the date of application.

(e) The delinquency fee at not more than 50 percent of the renewal fee in effect on the date of reinstatement.

(f) The board shall establish by regulation an appeal fee for examination. The regulation shall include provisions for an applicant to be reimbursed the appeal fee if the appeal results in passage of examination. The fee charged shall be no more than the costs incurred by the board.

(g) All other document fees are to be set by the board by rule.

Applicants wishing to be examined in more than one branch of engineering shall be required to pay the additional fee for each examination after the first.

{Amended by Stats.2007, Ch. 354}

PART THREE

Contracts

Los Angeles Air Force Base Environmental Assessment

Proposal 1234-05-011

PROPOSAL/WORK ACCEPTANCE SHEET

Project Name ("Project"): LAAFB Environmental Assessment

Proposal No. ("Proposal"): 1234-05-001

Date of Proposal: 6/16/2005

Project Location ("Site"): Los Angeles Air Force Base. CA

Office Designation: Los Angeles

CLIENT INFORMATION

Client Name: US Air Force

Attention: Major Jesus Peron

Address: 1234 Executive Ct LAAFB. CA 91111

E-mail Address: jperongilaafb.af.mil

Telephone Number: 310-985-1978

Fax Number: 310-985-1900

Scope of Authorized Services: Please refer to the attached Proposal, which is incorporated herein by reference.

This agreement ("Agreement") is effective this _____ day of _____200_ by and between CLIENT, with offices at the location set forth above, ("Client") and CONSULTANT, with offices at _____, ("CONSULTANT") (individually, "Party" and, collectively, "Parties"). In consideration of the mutual covenants expressed herein, and intending to be legally bound, the Parties agree as follows:

1. **STANDARD OF CARE.** CONSULTANT will perform the Scope of Authorized Services referenced above and in any subsequent change order ("Services") as an independent contractor, using that degree of skill and care ordinarily exercised under similar conditions by reputable members of CONSULTANT'S profession practicing in the same or similar locality at the time of performance. No other warranty, express or implied, is made or intended, and the same are specifically disclaimed, including the implied warranties of merchantability and fitness for a particular purpose. Any claim that Client may bring against CONSULTANT with respect to the Services to be performed by CONSULTANT must be commenced within one (1) year after the date on which Client first knew or should have known of the deficient Services upon which the claim is based.

2. OPINIONS OR ESTIMATES OF COST. If included in the Services, CONSULTANT will provide estimates of the cost of materials, remediation or construction based on CONSULTANT'S experience on similar projects, which are not intended for Client's or others' use in developing firm budgets or financial models, or making investment decisions. Client agrees that any opinion of cost is still merely an estimate.

3. PAYMENT. Client will pay CONSULTANT for performing the Services in accordance with the fee provisions contained in the Proposal or the above Scope of Authorized Services. CONSULTANT'S invoices are due upon receipt. If payment is not received within thirty (30) days from the date of CONSULTANT'S invoice, in addition to the invoiced amount, Client will owe interest on the unpaid balance at a rate equal to the lesser of one and one-half (1 $^1/_2$ %) percent per month or the maximum legal rate until the invoice, plus interest, is paid in full. CONSULTANT may suspend the performance of Services when any invoice is past due.

4. OWNERSHIP OF DELIVERABLES AND RELATED DOCUMENTS. All reports, drawings, plans, designs and other documents prepared by CONSULTANT pursuant to this Agreement, (collectively, "Deliverables") including all intellectual property rights therein, remain the property of CONSULTANT.

Client agrees that the Deliverables (i) are intended for the exclusive use and benefit of, and may be relied upon only by, Client and (ii) will not be used at a Site or for a Project not expressly provided for in this Agreement. Client does not require CONSULTANT'S permission for regulatory submittal of the Deliverables or, subject to all terms and conditions contained in this Agreement, reliance on the Deliverables provided to Client's design team, of which CONSULTANT is a member, solely for the design of the Project for which the Deliverables were intended. Client agrees that any prospective lender, buyer, seller or other third party who wishes to rely on any Deliverables must first sign CONSULTANT'S Secondary Client Agreement. Client will defend, indemnify and hold harmless CONSULTANT, its parent, subsidiaries, affiliates and subcontractors, including their respective officers, directors, employees, agents, successors and assigns, (collectively, "Indemnitees") from and against all claims, settlements, costs, expenses, liabilities, damages, penalties and interest, including attorneys' fees and litigation expenses, (collectively, "Liabilities") asserted against or incurred by Indemnitees as a result of any unauthorized use of, or reliance on, the Deliverables. If Client requests Deliverables on electronic media, it agrees that the electronic copy may be inaccurate or incomplete, and the document retained by CONSULTANT will be the official document. Any modification(s) of the electronic copy by Client will be at its own risk. Client further agrees that (i) the Deliverables may be based in part or in whole on facts and/or assumptions provided to, but not independently verified by, CONSULTANT, (ii) the Deliverables will reflect CONSULTANT'S findings as to conditions that existed only at the time the Services were performed and (iii) CONSULTANT makes no representations as to any facts or assumptions provided to, but not independently verified by, CONSULTANT.

5. LIMITATION OF LIABILITY. To the maximum extent permitted by law, client expressly agrees, for itself and anyone claiming by, through or under it, that the liability of consultant, its parent, subsidiaries, affiliates and subcontractors, including their respective officers, directors, employees, successors and assigns, for any and all causes of action whatsoever, including, without limitation, tort, contract, strict liability, indemnity or otherwise, arising out of, or in connection with, this agreement or consultant's professional services, shall be limited to the aggregate sum, including amounts paid to third party recipients of deliverables, attorneys' fees and all other litigation costs and expenses, if any, of fifty thousand dollars ($50,000) or the total fees paid to consultant by client under this agreement, whichever is greater.

Client acknowledges that (i) without the inclusion of this limitation of liability provision, CONSULTANT would not have performed the Services, (ii) it has had the opportunity to negotiate the terms of this limitation of liability as part of an "arms-length" transaction, (iii) the limitation amount may differ from the amount of professional liability insurance required of CONSULTANT under this Agreement, (iv) the limitation of liability provision is merely a limitation of, and not an exculpation from, CONSULTANT'S liability and (v) it has received special consideration of ten dollars ($10) for this limitation of liability provision and waives any and all rights to dispute the receipt and sufficiency of such consideration.

6. **INDIRECT DAMAGES.** Each party hereby waives its rights to recover from the other party any consequential, indirect, or incidental damages (including, but not limited to, loss of use, income, profits, financing or reputation), arising out of, or relating to, this agreement or the performance of the services, or both.

7. **SITE OPERATIONS**. If sampling or intrusive services are part of the Services, the Proposal or opinion of costs does not include the costs associated with surveying the Site to determine accurate horizontal and vertical locations of any tests, borings, or well installation locations. Client will establish test or boring locations. If surveying is required, those Services will be secured by Client. Field tests or boring locations described in the Deliverables or shown on sketches are based on information furnished by others or estimates made in the field by CONSULTANT'S personnel. Such depths, dimensions, or elevations are approximations. Unless expressly stated otherwise, the Services do not include the costs of restoration of damage which is reasonably necessary to perform the Services. Client will defend, indemnify and hold harmless the Indemnitees from and against Liabilities related to, or arising from, any undisclosed or unknown surface or subsurface conditions, except to the extent such Liabilities were caused solely by the negligence of the Indemnitees.

8. **SITE RESPONSIBILITY.** Client will provide CONSULTANT with access to the Site and all available Site Information deemed necessary by CONSULTANT. The Services do not include supervision or direction of the means, methods or actual work of other consultants, contractors and subcontractors not retained by CONSULTANT. Client agrees that each such other party will be solely responsible for its working conditions and safety on the Site. CONSULTANT'S monitoring of the procedures of any such other party is not intended to include a review of the adequacy of its safety measures. It is agreed that CONSULTANT is not responsible for safety or security at the Site, other than for CONSULTANT'S employees, and that CONSULTANT does not have the right or duty to stop the work of others.

9. **SAMPLES AND WASTES.** Samples are generally consumed or altered during testing and are disposed of immediately upon completion of the tests. If Client directs CONSULTANT to retain any samples, CONSULTANT will take reasonable steps to retain them, at Client's expense, but only for a mutually acceptable time. CONSULTANT reserves the right to refuse storage of any samples. If the samples or wastes resulting from the Services or any soils or materials contain asbestos, molds, fungi, bacteria, viruses, or any other hazardous, radioactive or toxic substances, pollutants, and/or their constituents (collectively, "Contaminants"), CONSULTANT, at Client's direction and expense, will either (i) return such samples, wastes, soils or materials to, or leave them with, Client for appropriate disposal or (ii) using a manifest signed by Client as generator and arranger, transport such samples, wastes, soils or materials to an approved facility selected by Client for final disposal, using a transporter selected by Client. In so doing, CONSULTANT will be acting solely as an independent contractor for Client and will at no time assume title, constructive or express, to any such samples, wastes, soils, or materials. Client will defend, indemnify and hold harmless the Indemnitees from and against all Liabilities arising from the Indemnitees handling of such samples, wastes, soils or materials, except to the extent of CONSULTANT'S negligence or willful violation of any applicable law.

10. **UNANTICIPATED CONDITIONS.** Client will inform CONSULTANT in writing of all known Contaminants or other conditions existing on or near the Site that present a potential danger to health, the environment, or CONSULTANT'S equipment or personnel prior to commencement of the Services. Should

CONSULTANT encounter such conditions which were not reasonably anticipated or which increase the risk or cost, or both, involved in CONSULTANT'S performance of the Services, upon notice to Client, CONSULTANT, in its sole discretion, may (i) suspend the performance of Services and submit a change order to be signed by Client prior to proceeding or (ii) discontinue the performance of Services and terminate this Agreement. If the unanticipated condition presents an immediate or potential threat to health, safety, the environment, or CONSULTANT'S equipment or personnel, CONSULTANT will immediately inform Client, so that Client can notify the appropriate government authorities. If Client fails to do so, Client will hold CONSULTANT harmless if CONSULTANT provides such notice.

11. **EXCUSABLE DELAY.** CONSULTANT will not be in breach of this Agreement due to any delay or failure to perform any obligation pursuant to a schedule, if such delay or schedule failure results from circumstances beyond the control of CONSULTANT. In the event of any such delay, CONSULTANT will be entitled to an extension of the time to put performance of the Services back on schedule, and CONSULTANT will be compensated for any necessary and reasonable increased costs of performance that result from such delay.

12. **ENVIRONMENTAL INDEMNITY.** Client (i) expressly releases the Indemnitees from all Liabilities arising from, or related to, any exposure, release, or dispersal of Contaminants on or about the Site and (ii) will defend, indemnify, and hold harmless the Indemnitees from and against all such Liabilities arising from or caused by Contaminants ("Environmental Liabilities"), except to the extent that such Environmental Liabilities are determined to have been caused solely by the negligence of, or the willful violation of any applicable environmental health or safety law, by CONSULTANT.

13. **TERM AND TERMINATION.** The term of this Agreement shall commence on the month, day and year first written above and shall continue in effect until completion of the Services and final payment, unless earlier terminated by Client for its convenience upon fourteen (14) days prior written notice to CONSULTANT. In such event, Client will take possession of the Site and the materials and equipment thereon, provided that such materials and equipment have been paid for by Client. CONSULTANT will be paid for all Services performed up to the effective date of termination, plus reasonable demobilization expenses and expenses related to the cancellation of previously-placed orders and other commitments regarding the Project.

14. **DISPUTE RESOLUTION.** The Parties shall undertake in good faith to settle or compromise all disputes, controversies, or differences between them that arise out of, or are related to, the performance of a Party under this Agreement (individually, "Dispute" and, collectively, "Disputes") by means of amicable discussions. All Disputes shall be dealt with as follows:

Any time there is a Dispute, either Party may send a written notice to the other Party setting forth a description of the Dispute ("Notice of Dispute"). If the Dispute is not resolved during the first fourteen (14) days following receipt of the Notice of Dispute, either Party may seek to have the Dispute resolved by non-binding mediation pursuant to the construction industry rules of the American Arbitration Association. Promptly upon selection of a mediator, the Parties shall provide the mediator with copies of the Notice of Dispute, all related, relevant documents and a statement of their respective positions and shall request that the mediator meet with the Parties within twenty (20) days of such selection to consider and propose a resolution or a procedure for reaching a resolution.

If the Parties have not resolved the Dispute or have not agreed in a writing signed by an officer of both Parties to resolve the Dispute by binding arbitration, either Party, after sixty (60) days following receipt of the Notice of Dispute (regardless of whether any mediation process has occurred or is ongoing or concluded), may seek a resolution in any state or federal court that has jurisdiction over the Parties and the subject matter of the Dispute ("Court") Either Party may apply to a Court for an order, if necessary, granting preliminary relief to maintain the status quo, to avoid irreparable injury, or to obtain other emergency relief at any time during the process described above. Despite such application, the Parties will continue to participate in good faith in the procedures specified in this Section 14.

The procedures specified in this Section 14. shall be the sole and exclusive procedures for the resolution of Disputes. THE PARTIES HEREBY WAIVE TRIAL BY JURY WITH RESPECT TO ANY ACTION OR PROCEEDING BROUGHT IN CONNECTION WITH THIS AGREEMENT.

15. **INSURANCE.** CONSULTANT maintains the following insurance coverages:
a. Workers' Compensation Insurance - statutory amount.
b. Commercial General Liability Insurance - $1,000,000 per occurrence/$2,000,000 aggregate.
c. Automobile Liability Insurance - $1,000,000 combined single limit.
d. Professional Errors & Omissions - $1,000,000 per claim/$ 1,000,000 aggregate.

16. **ASSIGNMENT.** This Agreement may not be assigned by either Party without the written consent of the other Party, which said such consent shall not be unreasonably withheld or delayed. Client acknowledges that CONSULTANT may subcontract portions of the Services to its affiliated companies and/or utilize employees of its affiliated companies in performing the Services, without the approval of Client.

17. **SURVIVAL.** All of Client's and CONSULTANT'S obligations and liabilities, including, but not limited to, Client's defense and indemnification obligations and the limitation of liability provision in Section 5, and CONSULTANT'S rights and remedies with respect thereto, shall survive completion of the Services and the expiration or termination of this Agreement.

18. **SEVERABILITY.** If any provision of this Agreement is deemed invalid or unenforceable, it is the intent of the Parties that this entire Agreement not be invalidated or rendered unenforceable, that the remaining provisions shall continue in full force and effect and the invalid or unenforceable provision shall be interpreted and enforced as closely as possible to the intent of the Parties, or deleted if a valid or enforceable interpretation is not possible under applicable law, and that the rights and obligations of the Parties shall be construed and enforced accordingly.

19. **NO CONSTRUCTION AGAINST THE DRAFTER.** Each of the Parties has had an opportunity to negotiate the terms and conditions expressed herein; therefore, this Agreement will not be construed more strictly against either Party as the drafter.

20. **INTEGRATION.** This Agreement and other documents, if attached as exhibits hereto, constitute the entire Agreement between the Parties and supersede any previous written or oral contracts or negotiations. This Agreement and the above Scope of Authorized Services can only be changed by a written instrument signed by both Parties.

21. **GOVERNING** LAW. This Agreement shall be governed by the laws of the State of Georgia, without giving effect to its choice of law principles.

Signed this _____ day of _____ 200__   Signed this _____ day of _____ 200__

_____   _____
Signature of Authorized CONSULTANT Representative   Signature of Authorized Client Representative
Print Name and Title   Print Name and Title

PROPOSAL #1234-05-001 FOR ENVIRONMENTAL ASSESSMENT (EA) FOR THE CONSTRUCTION OF HAZARDOUS WASTE/MATERIAL STORAGE BUILDING AT FORT MACARTHUR AND AREA B LOS ANGELES AIR FORCE BASE, CA

BACKGROUND

The U.S. Air Force has determined that completion of an environmental impact analysis is required for the demolition associated with the demolition of construction of the new hazardous waste/material storage buildings at Fort MacArthur and Area B at Los Angeles Air Force Base (AFB), California. For the purposes of this project, and in accordance with applicable regulations, this impact analysis shall hereafter be known as the Environmental Assessment (EA).

The Environmental Impact Analysis Process (EIAP) is the process by which federal agencies facilitate compliance with environmental regulations. The primary legislation affecting these agencies' decision-making process is the National Environmental Policy Act (NEPA) of 1969, which the Air Force implements under Air Force Instruction (AFI) 32-7061, the *Environmental Impact Analysis Process,* as promulgated in 32 Code of Federal Regulations (CFR) Part 989. In accordance with NEPA, federal agencies are required to take into consideration potential environmental consequences of proposed actions in their decision-making process. The intent of NEPA is to protect, restore, or enhance the environment through well-informed federal decisions. The federally appointed Council on Environmental Quality (CEQ) was established under NEPA to implement and oversee federal policy in this process. The CEQ subsequently issued Regulations for Implementing the *Procedural Provisions of the National Environmental Policy Act* (40 CFR sections 1500-1508). These regulations specify that an Environmental Assessment be prepared to:
- Briefly provide sufficient analysis and evidence for determining whether to prepare an Environmental Impact Statement (EIS) or a finding of no significant impact (FONSI);
- Aid in an agency's compliance with NEPA when no EIS is necessary; and
- Facilitate preparation of an EIS when one is necessary.

To comply with NEPA and other pertinent environmental requirements—such as the Endangered Species Act, the National Historic Preservation Act, the Clean Air Act, etc.—and to assess impacts on the environment, the decision-making process would include a study of environmental issues related to the demolition of construction of the new hazardous waste/material storage buildings at Fort MacArthur and Area B.

The A/E consultant will utilize staff with Department of Defense (DoD) experience, including those with U.S. Air Force (USAF)-specific experience. This proposal is broken down into the following elements:

STATEMENT OF WORK

The following approach (by task) is proposed to be used in preparing an EA for demolition associated with the demolition of construction of the new hazardous waste/material storage buildings at Fort MacArthur and Area B at Los Angeles AFB. The proposed project approach is based upon work conducted at various USAF and other DoD-component bases pursuant to the procedural provisions and content of typical NEPA analyses of proposed actions of a similar nature. The project team includes staff with ongoing, recent, and relevant experience regarding NEPA requirements for DoD facilities.

It is A/E Consultant's understanding that the Air Force proposes to perform some preparatory demolition and then construct and operate a new hazardous waste/material storage buildings at Fort MacArthur and Area B at Los Angeles AFB.

The Fort MacArthur, Pacific Crest, and Pacific Heights MFH areas are located in the San Pedro community within the City of Los Angeles. The majority of Fort MacArthur is designated as an historic district. The Pacific Heights housing area comprises solely MFH and is located adjacent to retail development and the recently established White Point Nature Preserve. The Pacific Crest housing area comprises solely MFH and is surrounded by retail and non-military residential development.

Preparation of the EA shall consist of collecting, reviewing, and evaluating data to define baseline environmental conditions related to such factors as air quality, environmental noise conditions, socioeconomic factors, air and ground traffic, land use, coastal zone management, and natural resources. In the context of these baseline conditions, the EA shall assess potential environmental consequences resulting from the demolition of construction of the new hazardous waste/material storage buildings at Fort MacArthur and Area B. Secondary impacts such as possible pollutant emissions, fugitive dust, storm water runoff, waste generation, possible fuel spills, and socioeconomic impacts shall also be assessed. The A/E Consultant will visit the two sites. A/E Consultant will coordinate with representatives from the base and the Air Force Center for Environmental Excellence (AFCEE) prior to the site visit in an effort to obtain as much information as practicable that is specific to the proposed action and the affected environment. This information should facilitate discussions at the end of the site visit that will further define relevant topics and strategies for preparing the draft Description of Proposed Action and Alternatives (DOPAA).

The A/E Consultant shall coordinate (see task-specific assumptions) with federal, state, county, and local offices maintaining noise, air quality, ecological, and general environmental data to obtain copies of documentation relating to the site environment. The A/E Consultant shall also coordinate with base organizations to gather data, assess potential construction impacts, and determine measures to minimize construction impacts. The Air Force will assist in the identification of these organizations. At a minimum, the requirements of Executive Order 12372 and AFI 32-7060, *Interagency and Intergovernmental Coordination for Environmental Planning* (IICEP), shall be met.

The A/E Consultant shall visit the base to conduct interviews and collect data. The A/E Consultant shall perform a search of available literature for the purpose of obtaining relevant information on operations, existing conditions, and those resources most susceptible to environmental impacts. Pertinent data shall be analyzed using appropriate techniques and an interdisciplinary team of experts. The analyses shall include all necessary assessment and, wherever possible, shall include existing analyses if they are properly validated and verified. Where data gaps exist, a reasonably expected worst-case analysis shall be performed, and identified as such in the text. Current anticipated tasks included for successful completion of this project (the Environmental Assessment) are as follows:

1. **Kickoff Meeting/Work Plan/Site Visit.** The A/E Consultant will arrange for a kickoff meeting to review project objectives, goals, schedule, and anticipated deliverables. The A/E Consultant will determine what information is available at the installation including available mapping (GIS/electronic format preferred), available special studies, an inventory and details of any known or suspected contaminated sites, specifics regarding the Project design and/or other construction project details, and a detailed statement of the military mission and the intended purpose and/or justification for the proposed project. The A/E Consultant will obtain and review as much information as possible prior to the site visit, compare against applicable regulations and Los Angeles AFB requirements, adjust site visit checklists as appropriate, and finalize logistical details. The A/E Consultant will prepare and provide meeting minutes that will serve as a Work Plan for the remainder of the project.

Immediately after the kickoff meeting on the same day, A/E Consultant staff will conduct a site visit and review of Los Angeles AFB project sites, and in particular the areas proposed for demolition and project construction to assess the nature and magnitude of any potential environmental effects. This assessment will focus on anticipated effects of demolition and construction activities associated with implementation of the Project as well as anticipated effects of long-term operation and management of the system. The consultant will identify and document any data gaps for preparation of the EA.

2. **Administrative Draft EA.** Preparation of a DOPAA and project analyses, including additional data collection and analysis of potential environmental impacts on applicable resource areas, will be presented to the Air Force for internal review and comment in the form of an Administrative Draft EA. The Administrative Draft EA will incorporate the DOPAA and will be prepared in accordance with Air Force Instruction (AFI) 32-7061. Format and number of copies will be consistent with CDRL requirements, as applicable.

3. **Draft EA.** One conference call or over-the-shoulder review to discuss results of Administrative Draft EA is recommended. Upon completion of the Administrative Draft EA review, A/E Consultant personnel will respond to comments and incorporate corrections into a Draft EA. The Draft EA will be distributed for agency and public review as part of the IICEP process. Format and number of copies will be consistent with CDRL requirements, as applicable.

4. **Public Comments and Responses/Draft FONSI.** Seven (7) days following the close of the comment period, a Comment and Response Matrix will be submitted. Further, provided the Draft EA has not identified significant environmental impacts, a Draft Finding of No Significant Impact (FONSI) will be submitted.

5. **Final Environmental Assessment/FONSI Preparation.** A/E Consultant staff will incorporate comments and prepare a Final EA/FONSI. Fourteen (14) day after receiving a signed FONSI, the Final EA/FONSI will be submitted for distribution and publication.

6. **Electronic Deliverables.** The A/E Consultant will provide electronic deliverables in the appropriate format (Word, Excel, PDF, GIS, etc.) such that Los Angeles AFB personnel can easily utilize the material in the future. The preferred format for these deliverables will be identified during the site visit and based on interviews with Los Angeles AFB staff during the kickoff meeting.

7. **Project/Program Management.** Project and Program Management activities are proposed to be consistent with typical AFCEE requirements, including project kickoff conference call minutes documentation, budget and schedule analysis, invoicing, QA/QC on deliverables, etc.

FEE ESTIMATE

Task #	Key Task-Specific Assumptions	Total
	Cost Proposal Preparation	
1 Kickoff Meeting / Work Plan / Site Visit/Tour	Kickoff meeting to be held in person at LAAFB, with no more than three A/E Consultant representatives. Meeting to be held in morning. A site visit/tour and onsite plan and document review will be performed immediately after and on the same day as the kickoff meeting. The onsite time for this task will be no more than one day. It is assumed that the Work Plan will be no more than 2 – 3 pages.	$7,800
2 Prepare Administrative Draft EA	One EA will be prepared covering the construction at Fort MacArthur and at Area B. Environmental setting information for Area B is readily available from other existing EAs. Environmental setting information for Fort MacArthur will be obtained from existing documents other than previous EAs (as there are none) and from interviews with LAAFB personnel. The key assumption is that there will be minimal impact analysis for these two projects because the project consists of simply replacing metal sheds/conex units with permanent buildings in the same or adjacent locations; the activities for these units will remain the same as past and current activities; there will be no increase in traffic; no change in traffic patterns, the project sites are already paved/disturbed areas; there will be no heavy construction equipment that will be used that will contribute significant air emissions; and LAAFB has already obtained all required environmental permits except for SHPO approval at Fort MacArthur. The primary issue at Fort MacArthur will be replacement of a building (Building 78) within the historical district. Cultural resources information will be obtained from the LAAFB cultural resources personnel. All other issues are assumed to be very minor in the EA analysis discussion. The primary issue at Area B will be air quality, in that because of much other construction going on at the facility, they are close to the AQ limits. However, because of the very minor amount of emissions expected from this project, it is assumed that the limits will not be exceeded so a full qualitative conformity and impact analysis along with extensive discussion with SCAQMD will not be required. Lead-based paint and asbestos surveys and other original quantitative data collection or analyses will not be performed. Any discussions or coordination with other agencies will be done through telephone calls, e-mails, or letters rather than in-person meetings. If A/E Consultant solicits feedback from agencies, only two request attempts will be made. There will be no additional onsite meetings or onsite data collection other than what was accomplished in Task 1.	$13,500
3 Prepare Draft EA	Comments from LAAFB will be collected through e-mail, letters, and telephone conversation; there will be no onsite meetings to discuss or review LAAFB comments. It is assumed that there will be no more than a total of 8 man-hours needed to respond to and incorporate comments from LAAFB on the Admin Draft EA to prepare the Draft EA.	$2,800
4 Prepare Response to Comments/Draft FONSI	It is assumed that there will be no more than a total of 4 public/agency comments (not 4 letters, but 4 actual comments) that will need responding to, and/or no more than 16 hours to respond to comments. There will be no onsite meetings or any other meetings such as with agencies associated with this task.	$5,000
5 Prepare Final EA/FONSI	It is assumed that there will be no more than a total of 8 man-hours needed to respond to and incorporate comments from LAAFB on the Admin Draft EA to prepare the Draft EA. There will be no onsite meetings or any other meetings such as with agencies associated with this task.	$3,000
6 Electronic Deliverables		$900
7 Program and Project Management	It is assumed the project will have a duration of three months.	$2,000
GRAND TOTAL		$35,000
	Estimate depends upon site conditions and availability of preparation materials, location, and availability of background data (electronic maps, records, etc.). Please note that combining similar projects into one delivery order could result in reducing fees through implementation of Site Visit and Project and Program Management efficiencies.	

OTHER ASSUMPTIONS

Key assumptions for each specific task were listed in the table above. Other assumptions in developing a fixed-cost proposal are as follows:

- In Statement Of Work For Environmental Assessment For The Construction Of The New Hazardous Waste/Material Storage Building At Fort Macarthur And Area B, (5) Scope of Work, (2) Environmental Assessment Report, Item (h) says, "Obtain approval from regulatory agencies for construction of the facility." LAAFB has already obtained all required permits except for approval by the State Historic Preservation Office (SHPO). Therefore, it is assumed that the only task A/E Consultant will perform to meet this item is preparation of a letter to SHPO requesting its approval of the project at Fort Mac Arthur. This letter will be prepared by A/E Consultant cultural resources staff and reviewed and approved by a California-certified architectural historian.

- It is assumed that the subject historic building at Fort Mac Arthur has been recorded to California SHPO standards and has been evaluated as either a significant or not significant resource; the A/E Consultant will not record the building with SHPO, but will only prepare a letter to SHPO requesting its approval of the proposed action at Fort Mac Arthur.

- Base will provide access to proposed project areas (and will provide a support person to accompany as deemed appropriate for high security areas, if applicable).

- Electronic (preferably GIS) data for site property, buildings, master planning, traffic and roads, etc. exists and will be made available to the A/E Consultant.

- Any and all planning documentation related to the proposed The Project (e.g., facilities requiring demolition, anticipated duration of construction, infrastructure required for operation, operation and maintenance procedures, etc.) will be made available to the A/E Consultant (no independent research regarding the system is anticipated to be made by the A/E Consultant).

- Based on initial evaluation of the proposed action, the following resource areas are likely to experience negligible environmental impacts: geological resources, water resources, transportation/traffic, hazardous materials and wastes, air quality, socioeconomics, environmental justice, public health and safety, biological resources, land use, noise, and visual resources/aesthetics.

- Based on initial evaluation of the proposed action, the following resource areas are likely to require environmental evaluations performed by discipline specialists: air quality and cultural/historic resources.

- Input for Interagency and Intergovernmental Coordination for Environmental Planning (IICEP) and public notification process (e.g., mailing lists, letter and notification templates, agency consultation, etc.) will be provided by the A/E Consultant; this proposal does not include newspaper advertisement fees associated with public notification.

Bellevue Youth Theater Expansion

RFP #07-123
Bellevue Youth Theatre Expansion

City of
Bellevue

Proposal Information:		Submit Proposals To:
Proposal Number:	RFP #07-123	City of Bellevue Service First Desk 1st floor Attn: Contracting Services 450 – 110th Ave NE Bellevue, WA 98009-9012
Proposal Name:	Bellevue Youth Theatre Expansion	
Date Issued:	May 15, 2007	
Contact Person:	Ken Kroeger, Project Manager	
Email:	kkroeger@bellevuewa.gov	CLEARLY MARK ENVELOPE WITH "RFP #07-123"
Proposals Due:	**May 31, 2007 2:00 PM**	

General Information: The City of Bellevue (City) is located three miles east of Seattle, between Lake Washington and Lake Sammamish, and about ten miles west of the foothills of the Cascade Mountains. The City's resident population of ~117,000 and daily workforce of ~121,000 make it Washington's fifth-largest city. Bellevue is a prosperous, increasingly diverse city that has evolved from a "bedroom community" into the economic and cultural hub of the Seattle area's Eastside. The City has developed its downtown core into a major business and retail center while maintaining the safe, comfortable family neighborhoods for which it has long been popular.

The Parks & Community Services Department is dedicated to providing a healthy community through an integrated system of exceptional parks, open space, recreation, cultural and human services programs. Bellevue's nationally-accredited Parks & Community Services Department won the 2005 National Gold Medal Award for Excellence in Park and Recreation Management.

NOTICE: Notice is hereby given that proposals will be received by the City of Bellevue, Washington, for **RFP #07-123: Bellevue Youth Theatre Expansion** by filing with the City at the above location.

PURPOSE: Bellevue Parks & Community Services Department (Parks) is soliciting requests for proposals from qualified firms to provide design and engineering services for the renovation and expansion of the Bellevue Youth Theatre (BYT).

BRIEF SCOPE OF SERVICES (Additional Detail listed in Attachment "A"): The BYT project will require the consulting team to work closely with City Staff and the BYT Advisory Board in order to complete the program analysis, verify space requirements, and develop potential phasing strategies. The project will include schematic, design development, construction documentation, bidding, and construction administration phases. In addition, a public outreach process to solicit and distribute information to the immediate neighborhood and greater community at large will need to be included.

QUALIFICATIONS: The proposed Scope of Work to include **all work** needed to fully address the design, permitting, marketing, and construction of the Bellevue Youth Theatre renovation and expansion. Team expertise shall include, but not be limited to:
- Knowledge of Theatre Design and Construction
- Public Outreach and Presentation(s)
- Marketing/Graphic Illustration
- Site and Building documentation/"as-built" (documentation of existing conditions)
- Architecture and Site Design
- Engineering (Civil, Mechanical, Electrical, Plumbing, Traffic, and Structural)
- Construction and Bid Documentation (including Cost and Construction Estimating)
- Knowledge of relevant Building and Site Codes
- Knowledge of COB and other associated Permit Requirements
- Knowledge of Public Bidding Requirements
- Value Engineering
- Construction Administration, including, Project Closeout and "As-Built" coordination

SUBMITTAL REQUIREMENTS: Three (3) bound-completed proposals, one (1) unbound completed proposal, and one (1) .pdf format electronic file (on a readable CD) must be received by the date and time listed. No more than 25 pages, double-sided, will be accepted. The City, at its discretion, may make additional copies of the proposal for the purpose of evaluation only. The original proposal will include original signatures, in ink, by authorized personnel, on all documents that require an authorized signature. Proposals shall be bound (8.5"x11") such that they lay flat when opened.

SIGNATURES: Proposals shall be signed by one of the legally authorized officers of said corporation. If awarded the contract, the Contract shall also be so executed.

QUESTIONS: Upon release of this RFP, all vendor communications should be directed in writing via e-mail to the Project Manager listed below. Unauthorized contact regarding this RFP with other City employees may result in disqualification. Any oral communications will be considered unofficial and non-binding on the City. **Questions should be e-mailed no later than 5pm, Thursday, May 24th.** Questions and answers will be sent electronically, and posted on the City's website no later than Tuesday, May 29th. The Project Manager for this RFP will be:

Name:	**Ken Kroeger—Project Manager**
Address:	City of Bellevue—Parks & Community Services Department
	450 – 110th Ave NE
	Bellevue, WA 98004
E-mail:	kkroeger@ bellevuewa.gov

EVALUATION CRITERIA & PROCESS: The objective is to select the vendor most qualified to provide the services outlined in this request. Evaluations will be based on the criteria listed below, which may be weighted as deemed appropriate by the review team/panel. All proposals will be evaluated using the same criteria:

Responsiveness, Quality, and Completeness of Proposal
- Are all the forms completed and everything included that was required by the RFP
- The quality of written material and presentation, relevant experience, answers to questions, and overall organization
- Clarity of responses

Experience/Qualifications/Creativity
- Experience working within the requested services arena
- Experience working with municipalities
- Ability to successfully complete the scope of services on time and within budget
- Ability to successfully work with City staff
- References
- Does the selected team have proven capability to deliver strong and innovative solutions that respond to key issues (Demonstrate capabilities in proposals)

Scope of Services
- Does the vendor understand what it will take to successfully achieve the goals and objectives of the requested services
- Did the vendor propose any revisions and/or changes to the Scope of Services that would better serve the City

Communication
- Vendor's ability to conduct public outreach meetings and solicit input
- Vendor's ability to communicate ideas, reports, and vision(s) in a clear and concise format
- Success of the project will rely on input from the community and from potential partners—how is this feedback incorporated into the process and final product

Qualifications of Key Individuals
- What personnel will be committed to this project and how are they uniquely suited for this project

After the proposals are evaluated, the City will determine whether formal presentations and interviews are necessary, and if so, which vendors may be invited to make a formal presentation and/or sit for a panel interview. The City may choose not to conduct formal presentations or interviews. The City may choose to contact officials from other jurisdictions regarding the vendor, their prior work experience and their ability to successfully complete the scope of services. The City may request clarification or additional information from a specific vendor in order to assist in the City's evaluation of a proposal. Finally, the City may require changes in the scope of services as deemed necessary by the City, before execution of the contract.

REJECTION OF PROPOSALS: The City reserves the right to reject any and all Proposals and to waive irregularities and informalities in the submittal and evaluation process. This RFP does not obligate the City to pay any costs incurred by respondents in the preparation and submission of their Proposals. Furthermore, the RFP does not obligate the city to accept or contract for any expressed or implied services.

CONTRACT AWARD: The City reserves the right to make an award without further discussion of the submittals. The Vendor selected as the apparently successful Vendor will be expected to enter into a contract with the City. A sample City contract is provided on the City's website at www.bellevuewa.gov.

To view a PDF version of the sample contract document click on "Find", then click "Bid Information" and select "Professional Services Contract". Please review this contract prior to submitting a Proposal.

Following consultant selection, the successful vendor shall prepare a proposal and scope of work for review by the City. Once the City and vendor have reached an agreement on the scope of services, a final contract will be prepared by the City. The foregoing should not be interpreted to prohibit either party from proposing additional contract terms and conditions during the negotiations of the final contract. If the selected vendor fails to sign the contract within ten (10) business days of delivery of the final contract, the City may elect to negotiate a contract with the next-highest ranked vendor. The City shall not be bound, or in any way obligated, until both parties have executed a contract. No party may incur any chargeable costs prior to the execution of the final contract.

The City reserves the right to award multiple contracts to multiple vendors for this scope of service if it is in the best interest of the City.

CONTRACT NEGOTIATION: The City reserves the right to negotiate all elements of the submittals, proposals, terms and conditions, and/or scope of services as part of the contract negotiation process prior to any formal authorization of the contract by the City. All parties understand that if any adjustments are made to the original scope as a result of contract negotiations, a resulting change in price/cost may be necessary.

EQUAL OPPORTUNITY EMPLOYMENT: The successful vendor or vendors must comply with the City of Bellevue equal opportunity requirements. The City of Bellevue is committed to a program of equal employment opportunity regardless of race, color, creed, sex, age, nationality or disability.

TITLE VI: It is the City of Bellevue's policy to assure that no person shall, on the grounds of race, color, national origin or sex, as provided by Title VI of the Civil Rights Act of 1964, be excluded from participation in, be denied the benefits of, or be otherwise discriminated against under any of its federally funded programs and activities.

INSURANCE REQUIREMENTS: The selected vendor or vendors shall maintain insurance that is sufficient to protect the Vendor's business against all applicable risks, as set forth in the City's Standard Insurance Requirements (attached to this document): **Attachment "B", plus the "Special Rider"** for professional liability. Please review insurance requirements prior to submitting your proposal. If selected vendor is unable to meet these standard requirements, please note current or proposed insurance coverages in submittal. Standard requirements may be negotiated if it is in the best interest of the City.

BUSINESS REGISTRATION AND TAXATION: The vendor or vendors awarded the contract will be subject to City of Bellevue Business Registration and Business Taxation as presented in the Bellevue City Code. Questions about the city's Business and Occupation (B&O) tax should be directed to the City's Tax office at 425-452-6851. The vendor awarded the contract will be subject to City of Bellevue business registration and business taxation as provide in Chapters 4.20 and 4.08 of the Bellevue City Code (for details call the City Tax office at 425-452-6851).

NON-ENDORSEMENT: As a result of the selection of a vendor to supply products and/or services to the City, vendor agrees to make no reference to the City in any literature, promotional material, brochures, sales presentation or the like without the express written consent of the City.

NON-COLLUSION: Submittal and signature of a proposal swears that the document is genuine and not a sham or collusive, and not made in the interest of any person not named, and that the vendor has not induced or solicited others to submit a sham offer, or to refrain from proposing.

COMPLIANCE WITH LAWS AND REGULATIONS: In addition to nondiscrimination and affirmative action compliance requirements previously listed, the vendor or vendors ultimately awarded a contract shall comply with federal, state and local laws, statutes and ordinances relative to the execution of the work. This requirement includes, but is not limited to, protection of public and employee safety and health; environmental protection; waste reduction and recycling; the protection of natural resources; permits; fees; taxes; and similar subjects.

PUBLIC RECORDS: Under Washington state law, the documents (including but not limited to written, printed, graphic, electronic, photographic or voice mail materials and/or transcriptions, recordings or reproductions thereof) submitted in response to this RFP (the "documents") become a public record upon submission to the city, subject to mandatory disclosure upon request by any person, unless the documents are exempted from public disclosure by a specific provision of law. If the city receives a request for inspection or copying of any such documents it will promptly notify the person submitting the documents to the city (by U.S. mail and by fax if the person has provided a fax number) and upon the written request of such person, received by the city within five (5) days of the mailing of such notice, will postpone disclosure of the documents for a reasonable period of time as permitted by law to enable such person to seek a court order prohibiting or conditioning the release of the documents. The City assumes no contractual obligation to enforce any exemption.

SUBMITTAL REQUIREMENTS: The following shall be addressed as a minimum, with sections tabbed with numbers as follows:
1. Introductory Letter:
   - Stating your interest in the project
   - How your firm's expertise is demonstrated by previous experience
   - A clear and concise response as to *why* your firm should be selected for this scope of work—recent professional experience, including work with public sector clients and applicable projects
2. Firm information:
   - Firm name, address, phone, fax, and contact information
   - Identify individual(s) your firm might assign as lead or project manager and a listing of the team members who will actually be assigned to perform substantial amounts of the work on this project, with a statement of their qualifications (resume and experience record for each person, including years of experience, education, and anticipated amount of time each will actually work on this project)
3. A minimum of three (3) local references with full name, title, address, email, phone, and fax number
4. Subconsultant experience:
   - Identify by discipline, any subconsultant firm(s)—Firm name, address, phone, fax, and
   - contact information
   - Include each proposed subconsultant(s)'s experience and qualifications as described
   - above for firm's personnel.
5. Briefly answer the following as part of the proposal:
   - What makes your firm uniquely qualified for this project?

- Include a statement of your design aesthetic or approach
- What values or benefits do you bring to the process that differentiates you from the other candidates?
- What experiences (good and bad) on other projects would help you make this project a success for everyone involved?
- Describe the process/system you would implement to produce a team or "partnering" relationship among those involved with the project, specifically with regard to interaction with Parks staff, the BYT Advisory Board, and the public.
- Identify/inform the project deliverables and task list for this project, beyond those listed in Attachment A.
- Include an overall project schedule for completion
- If you could have any "super power" what would it be? What would you use your power for? Why?

Submittals will not be returned.

ATTACHMENT A

PROJECT INFORMATION & SCOPE:

The proposed scope of work for the Bellevue Youth Theatre will require a value-added, budget conscious, transparent, and understandable design process that informs and coordinates programming, construction, maintenance, budget, and implementation timeline decisions.

BACKGROUND: Bellevue Youth Theatre at Ivanhoe Park

Bellevue Youth Theatre's mission is to provide opportunities in performing arts for all young people, regardless of income or ability, and allow these young people to perform before a live audience, to provide the community with quality productions which are entertaining and/or socially relevant for the entire family.

Bellevue Youth Theatre is a non-traditional recreation theatre program, uniquely designed to embrace young people from all backgrounds and abilities. Bellevue Youth Theatre puts every interested participant on stage. Every child is given an important role to play and no child is turned away—the youth and their experience are more important. This results in an environment that promotes self-esteem and is attractive to youth who may not have the financial means or ability to be involved in other theatre programs.

The Theatre attracts youth who want to be part of a quality theatre program while enjoying the company of their peers. The Theatre also provides a supportive environment for parents who want to remain involved in their children's activities. Many parents help with the technical phase of productions or even act with their children.

Bellevue Youth Theatre started as an outreach program at Crossroads Community Center in 1990. The program, originally named Crossroads Youth Theatre, grew with the addition of a multipurpose theatre space in 1992. In 2001, the City renovated the Ivanhoe Elementary School site for the 110 seat black box theatre. The renovation to the 6,100 SF facility included the main black box theatre space, restroom upgrade, lobby and staff/ticket office, mechanical spaces, green room, and other storage/miscellaneous spaces. The adjacent 3,500 SF 'annex' building currently houses Bellevue School District storage and a small BYT costume storage space.

In 2004, a new comprehensive business plan and a new name, Bellevue Youth Theatre was introduced, and along with the formation of the Bellevue Youth Theatre Academy, a new phase was begun for the BYT. However, the BYT has now reached its space capacity and the ability for the programs to grow is limited.

Park staff conducted a series of charettes with the BYT Board to identify the needs of the theatre. These charettes evaluated the existing theatre for current program areas, perceived successful and problematic areas, future growth options, and anticipated development costs. The initial findings determined that the BYT needs significant upgrades to improve safety and security and match Bellevue service standards, including: expanded restrooms, increased lobby space, additional secured dressing rooms (for various ages and gender), and improved parking and lighting. Additionally, the BYT needs to be expanded to meet the increasing demand for programs—more rehearsal and classroom space, secured staff and volunteer offices, as well as critical storage is needed (see Attachment D—Concept Plans and Program).

The BYT Foundation supports this project in order to create capacity to continue growth of this successful program, to protect the theatre's mission of inclusiveness, to bring the theatre up to standards expected of Bellevue Parks' facilities, to fulfill the BYT Business Plan strategy for expansion, and to demonstrate effectiveness of City/Community partnership in the long range vision for Bellevue. The expansion has been estimated to be $2.5 to $3.5 million and will allow the BYT to continue to serve the community for many years to come.

PUBLIC OUTREACH: A public involvement process will be required as part of the scope of work. The City will assist in the organization, however, the overall coordination and documentation will fall within the contracted scope of work.

The design team is expected to participate in a minimum of one (1) public meeting to relay general information to the community and solicit input, a minimum of two (2) meetings to inform and seek input from the BYT Board. Additionally, the team may be asked to present to the City's Park & Community Services Board. Any additional outreach meetings beyond this minimum will be determined by the City.

PROGRAMMING: This step will assist to address the opportunities, constraints, and benefits of the existing facility and how the newly identified program requirements will affect the current use. This will assist to verify square footage of the size and arrangement of the spaces: storage, instructional opportunities, recreational components, and other requirements. Additionally, the team will need to work with the City and BYT members to review and identify how the conceptual components of the BYT expansion will be structured:
- Potential programming conflicts (site and building) to be included for analysis
- Potential code issues and recommendations
- How the individual Programmatic elements effect the current programming and existing structures
- Schedule/timeline showing when major milestones for tasks and subtasks will be completed

PLANNING/DESIGN: The planning and design phases will include all diagramming and documentation, and will incorporate recommendations received from City staff, the community, and/or the BYT Board, through construction documentation. Initial building strategies have been identified, and will be used as a basis for design (see Attachment D—Concept Plans and Program).

This stage will need to identify the critical components of the BAC and address:
- How the individual programmatic elements effect the existing facility and site
- Code requirements and potential permit conflicts and constraints
- Cost estimating
- Marketing strategies and analysis, including various conceptual illustrations and marketing tools

Conceptual design documents will need to include, at a minimum: written program; site and building diagrams and graphic presentations containing design objectives, space requirements, relationships, site conditions, perspective sketches/renderings, computer modeling, and/or a combination of these media; potential traffic study and concerns; code requirements and potential conflicts; and preliminary construction and operational estimates. These estimates will be divided into specific program components/features which can be added or deleted from the conceptual complex.

Tasks associated with the development of the design include, but are not limited to:

Task 1 – Contract Scope
  Goal: Define and agree to a Scope of Work.

Approach: Draft Scope for review by the Parks Department. Final Scope of Work will incorporate Parks' comments and direction, and inform the Professional Services Contract.

Assumptions: Parks will review and provide input to complete the Final Scope of Work.

Deliverables: Draft and Final Scope of Work, and executed Professional Services Contract with insurance requirements.

Task 2 – Program Review and Development

Goal: Assist Parks and BYT Board in determining the appropriate programming, space, and relational qualities for optimum theatre programming and use.

Approach: Conduct on-site walk-through(s)/meeting(s) with Parks Staff to evaluate the site(s), and opportunities for the design(s).

Assumptions. Prepare site walk-through, meeting agenda and necessary data for review. Conduct and summarize the working session. No current complete as-built documents exist for the theatre or auxiliary building.

Deliverables: Recommendations for final program and site walk-through agenda and minutes. Complete documentation and survey of existing facility and site condition.

Task 3 – Schematic Design

Goal: Develop multiple schematic design options reflecting desired programming and potential budget. Include options for development and expansion to include: the demolition of the annex building; demo of existing theatre and annex; and those options identified from the BYT design charette.

Approach: Create appropriate design(s) for the theatre expansion, addressing prioritized amenities, cost constraints and coordination of existing conditions for an integrated layout. Participate in public outreach meeting(s) with Parks staff to evaluate the design.

Assumptions: Prepare site walk-through, agenda and necessary data, drawings for review. Attend an open house, conduct and summarize the working sessions.

Deliverables: Recommendations for final design, public meeting(s) agenda(s)/minutes, cost estimates to be submitted for review by COB and BYT at 50% and 90% completion of schematic design.

Task 4 – Design Development

Goal: Refine the preferred schematic design option reflecting the desired programming, as well as, working with the project budget constraints—develop and expand the schematic design documents relating to form, size, and appearance of the project. It includes specific structural, mechanical and electrical systems, and interior design. Provide options for material(s) selection and begin the development of the Project Specifications—Outline specification, itemized lists and identification of significant materials, systems, and equipment (CSI format with COB front end coordination). Review and update the previously established schedule. Refine previous SD estimate based on adjustments made during the design development phase.

Approach: Create an appropriate design for the theatre expansion, addressing prioritized amenities, cost constraints and coordination of existing conditions for an integrated layout. Participate in public outreach meeting(s), with Parks staff to evaluate the design. All design development services and documentation is presented to the owner for approval.

Assumptions: Prepare meeting(s) agenda and minutes, create and assemble DD drawings and specifications for review at 50% and 90% DD, and focus final design strategy.

Deliverables: Final design strategy documentation, CD and Project Manual template, public meeting(s) agenda(s)/minutes, and cost estimates for review by COB at 50% and 90% DD. Final graphic presentation(s) items are to include multiple renderings and 3D model of theatre for anticipated BYT fundraising campaign.

Task 5 – Construction and Bid Documentation

Goal: Create a complete construction and bid documentation package, Construction Drawings and Specifications (Summary of Work and Technical Specifications), and potential Value Engineering for the Bellevue Youth Theatre expansion. Coordinate with Parks Department to fully integrate City of Bellevue "Front-End" Specifications.

Approach: **TBD**

Assumptions: Parks to review plans and specifications at 50% and 90%. Parks to provide City of Bellevue standard "Front-End" specifications for inclusion into the contract documents.

Deliverables: Working meeting agenda(s) / minutes, permit and bid-ready plans and specifications provided to the City of Bellevue in hard copy and digital format, and cost estimates for 50 and 90% CDs.

Task 6 – Permit Process

Goal: Obtain all required permits and approvals for construction.

Approach: Submit the construction documents for permit review to the appropriate agencies. Work with the reviewing agencies and Parks to coordinate any permit comments and required corrections/ modifications.

Assumptions: Parks to assist in permit submittal, but is not the lead permit official, reviewer or coordinator. Permit fees to be paid by Parks.

Deliverables: All permit documents and revisions as required.

Task 7 – Bid Process

Goal: Assist in managing the bid process for the Bellevue Youth Theatre Expansion.

Assumptions: The City of Bellevue Contracting Services and Parks departments will take the lead for the bid, including reproduction of construction documentation.

Approach: Prepare and conduct site walk-through, including agenda and necessary data, drawings for review. Provide all bid clarification/addenda.

Deliverables: Site walk-through agenda/ minutes, and all bid documents and revisions/addenda as required. Bid Evaluation.

Task 8 – Construction Administration

Goal: Coordinate, document, and assist Parks Project Manager with the construction of the Bellevue Youth Theatre expansion

Approach: Problems are to be identified quickly and corrective action pursued with minimal delay, if any, to the overall project. Part of the information to be developed is weekly progress reports, and other correspondence, as appropriate.

Weekly report, addressing progress of the work, shall include (as appropriate):
- A summary of work to date
- A summary of actual versus scheduled progress
- A list and brief summary of any Change Order Proposals
- Digital photo documentation of construction
- Tracking and review of all Shop Drawing and other construction related
- submittals.

Coordinate project documentation, including the following:
- Prepare all necessary project correspondence, letters, memos, meeting minutes, etc., for support to the project work. Maintain a central file for all written materials.

Deliverables: Construction Documents, Weekly Progress Reports, change order proposal logs, shop drawing and submittal logs, as-Built drawings, and project correspondence (as required).

ATTACHMENT "B"

INSURANCE REQUIREMENTS

The Contractor shall procure and maintain for the duration of this Agreement insurance against claims for injuries to persons or damages to property which may arise from or in connection with the performance of the work hereunder by the Contractor, his agents, representatives, employees or subcontractors. The cost of such insurance shall be paid by the Contractor. Insurance shall meet or exceed the following unless otherwise approved by the City.

A. **Minimum Insurance**
1. Commercial General Liability coverage with limits not less than $1,000,000 per occurrence / $2,000,000 annual aggregate,
2. Stop Gap/Employers Liability coverage with limits not less than $1,000,000 per accident/disease,
3. Business Automobile Liability coverage with limits not less than $1,000,000 per accident for any auto,
4. Workers' Compensation coverage as required by the Industrial Insurance Laws of the State of Washington.

B. **Self-Insured Retentions**
Self-insured retentions must be declared to and approved by the City.

C. **Other Provisions**

Commercial General Liability policies shall be endorsed to:
1. Include the City, its officials, employees and volunteers as insureds,
2. Provide that such insurance shall be primary as respects any insurance or self-insurance maintained by the City,
3. Each insurance policy shall provide that coverage shall not be canceled except after thirty (30) days' written notice has been given to the City.

D. **Acceptability of Insurers**
Insurance shall be placed with insurers with a rating acceptable to the City.

E. **Verification of Coverage**
Contractor shall furnish the City with certificates of insurance required by this clause. The certificates are to be received and approved by the City before work commences. The City reserves the right to require complete, certified copies of all required insurance policies at any time.

F. **Subcontractors**
Contractor shall require subcontractors to provide coverage which complies with the requirements stated herein.

> Approved as to form:
>
> *Patricia C. Cole*
> Assistant City Attorney

J:\General Services\Confidential\ES\attachmentB.doc 9/30/2004

SPECIAL RIDER

TO

ATTACHMENT "B"

INSURANCE REQUIREMENTS

for

Professional Service Contracts

Add the following to section A. <u>Minimum Insurance:</u>

1. Consultant's Errors & Omissions or Professional Liability with limits not less than $1,000,000 per claim and as an annual aggregate.

CITY OF BELLEVUE
City Attorneys Office
425 452-6829

REC. NO._____
CITY OF BELLEVUE
DATE_____
_____
CITY CLERK'S OFFICE
CCO FILE #_____

City of Bellevue Contract Identification Information:

Contract Title: **Bellevue Youth Theatre Expansion** _____

PROFESSIONAL SERVICES AGREEMENT

THIS AGREEMENT is entered into the date last below written between the **CITY OF BELLEVUE, WASHINGTON ("CITY")** and_____
_____**("CONTRACTOR")**.

1. SERVICES BY CONTRACTOR

A. Performance of Services. The Contractor shall perform the services described in the scope of work attached hereto as Attachment A. All Services will be rendered using the highest professional standards and in a timely, first-class and professional manner, in compliance with all standards and rules reasonably established by the City.

B. Modification. The City periodically may make changes to the Services that are within the general scope of the Agreement, by giving the Contractor written notice of such changes. If any change results in an increase or a reduction in the work that was contemplated to be performed by the Contractor as described in Attachment A, the Contractor's compensation hereunder shall be modified accordingly.

2. PAYMENT

A. The City shall pay the Contractor for such services: *(Check One)*

 ○ Hourly: _____ per hour, plus actual expenses, but not more than a total of _____

 ○ Fixed Sum: A total amount of_____

 ○ Other:_____

for all services performed, and expenses incurred under this agreement.

B. The Contractor shall maintain time and expense records and provide them not more frequently than monthly to the City, along with monthly invoices in a format acceptable to the City for work performed to the date of the invoice.

C. All invoices shall be paid by mailing a city warrant within 30 days of receipt of a proper invoice after approval of the Contractor's completed tasks/deliverables to the date of the invoice or monthly report, as appropriate.

D. The Contractor shall keep cost records and accounts pertaining to the Agreement available for inspection by the City's representatives for three (3) years after final payment. Copies shall be made available on request.

E. If the services rendered do not meet the requirements of the Agreement, the Contractor will correct or modify the work to comply with the Agreement. The City may withhold payment for such work until the work meets the requirements of the Agreement.

3. DISCRIMINATION AND COMPLIANCE WITH LAWS

A. The Contractor agrees not to discriminate against any employee or applicant for employment or any other person in the performance of this Agreement because of race, creed, color, national origin, marital status, sex, age, disability, or other circumstance prohibited by federal, state or local law or ordinance, except for a bona fide occupational qualification.

B. The Contractor shall comply with all federal, state and local laws and ordinances applicable to the work to be done under this Agreement, including where applicable Bellevue City Code § 4.28.143.

C. Violation of this Section 3 shall be a material breach of this Agreement and grounds for cancellation, termination or suspension of the Agreement by the City, in whole or in part, and may result in ineligibility for further work for the City.

4. TERM AND TERMINATION OF AGREEMENT

A. Term. This Agreement shall remain in effect until completion of the services described in Attachment A and final payment therefor unless terminated earlier in accordance with Paragraph 4 of this Agreement.

B. Rights Upon Termination. This Agreement may be terminated by either party without cause upon thirty days' written notice, in which event all finished or unfinished documents, reports, or other material or work of Contractor pursuant to this Agreement shall be submitted to the City, and the Contractor shall be entitled to just and equitable compensation at the rate set forth in Paragraph 2 for any satisfactory work completed prior to the date of termination.

C. Noninterference with business. During the course of the Contractor's performance of the Services for the City and for a period of twelve (12) months after the completion of such Services, the Contractor will not interfere with the City's business in any manner, including without limitation, encouraging anyone to leave the City's employ or encouraging any employee or independent contractor to sever that person's relationship with the City.

5. OWNERSHIP OF WORK PRODUCT

All data, materials, reports, memoranda and other documents developed under this Agreement whether finished or not shall become the property of the City, shall be forwarded to the City at its request and may be used by the City as it sees fit. The City agrees that if it uses products prepared by the Contractor for purposes other than those intended in this Agreement, it does so at its sole risk and it agrees to hold the Contractor harmless therefor.

6. GENERAL ADMINISTRATION AND MANAGEMENT

The Director of the _____ Department of the City of Bellevue, or his/her designee, shall be the City's representative, and shall oversee and approve all services to the performed, coordinate all communications, and review and approve all invoices, under this Agreement.

7. INDEMNIFICATION AND HOLD HARMLESS

A. The Contractor shall protect, defend, indemnify and save harmless the City, its officers, employees and agents from any and all costs, claims, judgments or awards of damages, arising out of or in any way resulting from the negligent acts or omissions of the Contractor. The Contractor agrees that its obligations under this subparagraph extend to any claim, demand, and/or cause of action brought by, or on behalf of, any of its employees or agents. For this purpose, the Contractor, by mutual negotiation, hereby waives, as respects the City only, any immunity that would otherwise be available against such claims under the Industrial Insurance provisions of Title 51 RCW. In the event the City incurs any judgment, award, and/or cost arising therefrom including attorneys' fees to enforce the provisions of this article, all such fees, expenses and costs shall be recoverable from the Contractor.

B. The City shall protect, defend, indemnify and save harmless the Contractor, its officers, employees and agents from any and all costs, claims, judgments or awards of damages, arising out of or in any way resulting from the negligent acts or omissions of the City. The City agrees that its obligations under this subparagraph extend to any claim, demand, and/or cause of action brought by, or on behalf of, any of its employees or agents. For this purpose, the City, by mutual negotiation, hereby waives, as respects the Contractor only, any immunity that would otherwise be available against such claims under the Industrial Insurance provisions of Title 51 ROW. In the event the Contractor incurs any judgment, award, and/or cost arising therefrom including attorneys' fees to enforce the provisions of this article, all such fees, expenses and costs shall be recoverable from the City.

C. The Contractor will indemnify, defend, and hold the City (and its elected officials, officers, employees, successors, assigns, insurers, licensees, distributors, independent contractors, and agents) harmless from all claims, damages, losses, and expenses (including reasonable attorneys' fees incurred on such claims and in proving the right to indemnification) arising out of or resulting from any claim, action, or other proceeding that is based upon (a) the Contractor's breach of any obligations, representations, or warranties under the Agreement, (b) the Contractor's outside business activities, or (c) the infringement or misappropriation by the Contractor of any foreign or United States patent, copyright, trade secret, or other proprietary right in results.

8. INSURANCE; RISK OF LOSS

The Contractor shall maintain insurance that is sufficient to protect the Contractor's business against all applicable risks, as set forth in Attachment B. The Contractor will cause the indemnified parties, as described in Section 7, above, to be named as additional insureds on the policy required under the Agreement and shall cause its insurance to be primary to any insurance carried by the indemnified parties. The Contractor will provide the City with certificates of insurance and other supporting materials as City reasonably may request to evidence Contractor's continuing compliance with this Section 8. The Contractor will be liable for all loss or damage, other than ordinary wear and tear, to the City's property in the Contractor's possession or control that is cause by the Contractor. In the event of any such loss or damage, the Contractor will pay the City the full current replacement cost of such equipment or property within thirty (30) days after its loss or damage.

9. INDEPENDENT CONTRACTOR

A. Nature of Relationship. The Contractor shall be and act as an independent contractor (and not as the employee, agent, or representative of the City) in the performance of the Services for the City. The Agreement shall not be interpreted or construed as creating or evidencing an association, joint venture, partnership or franchise relationship among the parties or as imposing any partnership, franchise, obligation, or liability on any party. The Contractor will not represent himself/herself as an employee of the City. The Contractor shall not be entitled to, and shall not attempt to, create or assume any obligation, express or implied, on behalf of the City. So long as the Contractor is able to adequately perform all of the Contractor's obligations under the Agreement in a skilled and workmanlike manner, the Contractor shall not be required to devote the Contractor's full time to the performance of the Services called for under the Agreement, and it is acknowledged that the Contractor has other clients and/or offers services to the general public. Since the Contractor will not be an employee of the City, the Contractor will not be entitled to any of the benefits that the City may make available to its employees, such as but not limited to vacation leave, sick leave, or insurance programs, including group health insurance or retirement benefits; nor shall the Contractor permit or cause any of the Contractor's employees, agents or subcontractors to perform any services under the Agreement in such a way as to cause or enable them to become, or claim to have become, employees, common law or otherwise, of the City. In addition, the Contractor acknowledges that as an independent contractor, he/she/it and/or his/her/its agents, servants or employees are not eligible to recover worker's compensation benefits from or through the City in the event of injury.

B. Contractor Responsible for Taxes and Records. The Contractor will be solely responsible for and will file, on a timely basis, all tax returns and payments required to be filed with or made to any federal, state or local tax authority with respect to the Contractor's performance of the Services and receipt of fees under the Agreement. The Contractor will be solely responsible for and must maintain adequate records of expenses incurred in the course of performing the Services under the Agreement. No part of the Contractor's compensation will be subject to withholding by the City for the payment of any social security, federal, state or any other employee payroll taxes; nor shall the City be obligated to make any such withholdings and/or payments on behalf of any employee, subcontractor, supplier, or other person working for or engaged by the Contractor to perform the Contractor's obligations under the Agreement. The City will regularly report amounts paid to the Contractor by filing Form 1099-MISC with the Internal Revenue Service as required by law.

10. SUBLETTING OR ASSIGNING CONTRACT

Neither the City nor the Contractor shall assign, transfer, or encumber any rights, duties or interests accruing from this Agreement without the express prior written consent of the other.

11. FUTURE SUPPORT

The City makes no commitment and assumes no obligations for the support of the Contractor's activities except as set forth in this Agreement.

12. GENERAL PROVISIONS

A. Governing Law; Forum. The Agreement will be governed by the laws of Washington and its choice of law rules. The Contractor irrevocably consents to the exclusive personal jurisdiction and venue of the federal and state courts located in King County, Washington, with respect to any dispute arising out of or in connection with the Agreement, and agrees not to commence or prosecute any action or proceeding arising out of or in connection with the Agreement other than in the aforementioned courts.

B. Severability. If any provision of the Agreement is held to be invalid or unenforceable for any reason, the remaining provisions will continue in full force without being impaired or invalidated in any way. The City and the Contractor agree to replace any invalid provision with a valid provision that most closely approximates the intent and economic effect of the invalid provision.

C. Nonwaiver. Any failure by the City to enforce strict performance of any provision of the Agreement will not constitute a waiver of the City's right to subsequently enforce such provision or any other provision of the Agreement.

D. No Assignment. Neither the Agreement nor any of the rights or obligations of the Contractor arising under the Agreement may be assigned without the City's prior written consent. Subject to the foregoing, the Agreement will be binding upon, enforceable by, and inure to the benefit of, the parties and their successors and assigns.

E. City Marks. The Contractor will not use any trade name, trademark, service mark, or logo of the City (or any name, mark, or logo confusingly similar thereto) in any advertising, promotions, or otherwise, without the City's express prior written consent.

F. Notices. All notices and other communications under the Agreement must be in writing, and must be given by registered or certified mail, postage prepaid, or delivered by hand to the party to whom the communication is to be given, at its address set forth below.

G. Legal Fees. In any lawsuit between the parties with respect to the matters covered by the Agreement, the prevailing party will be entitled to receive its reasonable attorney's fees and costs incurred in the lawsuit, in addition to any other relief it may be awarded.

H. Counterparts. The Agreement may be signed in counterparts, each of which shall be deemed an original, and all of which, taken together, shall be deemed one and the same document.

13. EXTENT OF AGREEMENT/MODIFICATION

This Agreement, together with attachments or addenda, represents the entire and integrated Agreement between the parties hereto and supersedes all prior negotiations, representations, or agreements, either written or oral. This Agreement may be amended, modified or added to only by written instrument properly signed by both parties hereto.

IN WITNESS WHEREOF, the parties have executed this Agreement as of _____, 20____.

CONTRACTOR   **CITY OF BELLEVUE:**

By:_____   By:_____

Printed Name:_____   Printed Name:_____

Title:_____   Title:_____

Address:_____   Approved as to form:

City/State/ZIP:_____   By:_____
                                                              City Attorney

Tax ID #: _____

City of Rancho Palos Verdes Professional Services Agreement

PROFESSIONAL SERVICES AGREEMENT

THIS AGREEMENT is made and entered into this __ day of _____, by and between the City of Rancho Palos Verdes hereinafter referred to as "CITY", and_____, hereafter referred to as "CONSULTANT".

IN CONSIDERATION of the covenants hereinafter set forth, the parties hereto mutually agree as follows:

ARTICLE 1
SCOPE OF SERVICES

1.1 <u>Project Description</u>
2006 Water Quality and Flood Protection Program Engineering Design Professional Services

1.2 <u>Description of Services</u>
CONSULTANT shall perform Services described in CITY'S Request for Proposals (RFP) and the CONSULTANT'S Proposal, attached here as EXHIBIT "B".

1.3 <u>Schedule of Work</u>
Upon receipt of written Notice to Proceed from the CITY, CONSULTANT shall perform with due diligence the services requested by the CITY and agreed on by CONSULTANT. CONSULTANT shall not be responsible for delay, nor shall CONSULTANT be responsible for damages or be in default or deemed to be in default by reason of strikes, lockouts, accidents, or acts of God, or the failure of CITY to furnish timely information or to approve or disapprove CONSULTANT'S work promptly, or delay or faulty performance by CITY, other contractors, or governmental agencies, or any other delays beyond CONSULTANT'S control or without CONSULTANT'S fault.

ARTICLE 2
COMPENSATION

2.1 Fee

(a) CITY agrees to compensate CONSULTANT for services in accordance with the fee schedule that is attached here as Exhibit "A." The total fee shall not exceed _____ dollars, except as provided for in 2.1(b).

(b) CITY may request additional specified work under this agreement. All such work must be authorized in writing by the Director of Public Works prior to commencement.

2.2 Payment Address

All payments due CONSULTANT shall be paid to:
Name of the company
Address

2.3 Terms of Compensation

CONSULTANT will submit invoices monthly for the percentage of work completed in the previous month. CITY agrees to pay all undisputed invoice amounts within thirty (30) days of receipt of the invoice. CITY agrees to use its best efforts to notify CONSULTANT of any disputed invoice amounts or claimed completion percentages within ten (10) days of the receipt of each invoice. However, CITY's failure to timely notify CONSULTANT of a disputed amount of claimed completion percentage shall not be deemed a waiver of CITY's right to challenge such amount or percentage.

Additionally, in the event CITY fails to pay any undisputed amounts due CONSULTANT within forty-five (45) days after invoices are received by CITY then CITY agrees that CONSULTANT shall have the right to consider said default a total breach of this Agreement and be terminated by CONSULTANT without liability to CONSULTANT upon ten (10) working days advance written notice.

2.4 Additional Services

CITY may request in writing that CONSULTANT perform additional services not covered by the specific Scope of Work set forth in this Agreement, and CONSULTANT shall perform such services and will be paid for such additional services in accordance with CONSULTANT'S Standard Schedule of Hourly Rates.

ARTICLE 3
INDEMNIFICATION AND INSURANCE

3.1 Indemnification

CONSULTANT will defend, indemnify and hold harmless CITY, its Boards and its officers, employees and agents (collectively "CITY"), against any claim, loss or liability that arises because of the sole or primary negligence or willful misconduct of CONSULTANT, its agents, officers, directors or employees, in performing any of the services under this Agreement.

3.2 General Liability

CONSULTANT shall at all times during the term of the Agreement carry, maintain, and keep in full force and effect, a policy or policies of Commercial General Liability Insurance, with minimum limits of two Million ($2,000,000.00) Dollars for each occurrence and in the aggregate, combined single limit, against any personal injury, death, loss or damage resulting from the wrongful or negligent acts by CONSULTANT. Said policy or policies shall be issued by an insurer admitted to do business in the State of California and rated in Best's Insurance Guide with a rating of A VII or better.

3.3 Professional Liability

CONSULTANT shall at all times during the term of this Agreement, carry, maintain, and keep in full force and effect a policy or policies of professional liability insurance with a minimum limit of two million ($2,000,000.00) dollars. Said policy or policies shall be issued by an insurer admitted to do business in the State of California and rated in Best's Insurance Guide with a rating of A VII or better.

3.4 Worker's Compensation

CONSULTANT agrees to maintain in force at all times during the performance of work under this Agreement worker's compensation insurance as required by the law. CONSULTANT shall require any subcontractor similarly to provide such compensation insurance for their respective employees.

3.5 Notice of Cancellation

A. All insurance policies shall provide that the insurance coverage shall not be canceled by the insurance carrier without thirty-(30) day's prior written notice to CITY. CONSULTANT agrees that it will not cancel or reduce said insurance coverage.

B. CONSULTANT agrees that if it does not keep the aforesaid insurance in full force and effect, CITY may either immediately terminate this Agreement or, if insurance is available at a reasonable cost, CITY may take out the necessary insurance and pay, at CONSULTANT's expense, the premium thereon.

3.6 Certificate of Insurance

At all times during the term of this Agreement, CONSULTANT shall maintain on file with the CITY Clerk certificates of insurance showing that the aforesaid policies are in effect in the required amounts. The commercial general liability and professional liability policy or policies shall contain endorsements naming the CITY, its officers, agents and employees as additional insured.

3.7 Primary Coverage

The insurance provided by CONSULTANT shall be primary to any coverage available to city. The insurance policies (other than workers' compensation and professional liability) shall include provisions for waiver of subrogation.

ARTICLE 4
TERMINATION

4.1 Termination of Agreement

(a) This Agreement may be terminated at any time, with or without cause, by either party upon sixty-(60) day's prior written notice. Notice shall be deemed served upon deposit in the United States Mail of a certified or registered letter, postage prepaid, return receipt requested, addressed to the other party, or upon personal service of such notice to the other party, at the address set forth in Article 6.12.

(b) In the event of termination or cancellation of this Agreement by CONSULTANT or CITY, due to no fault or failure of performance by CONSULTANT, CONSULTANT shall be paid compensation for all services performed by CONSULTANT, in an amount to be determined as follows: for work done in accordance with all of the terms and provisions of this Agreement, CONSULTANT shall be paid an amount equal to the percentage of services performed prior to the effective date of termination or cancellation in accordance with the work items; provided, in no event shall the amount of money paid under the foregoing provisions of this paragraph exceed the amount which would have been paid to CONSULTANT for the full performance of the services described in Article 2.1.

ARTICLE 5
OWNERSHIP OF DOCUMENTS

5.1 Ownership of Documents and Work Product

All plans, specifications, reports and other design documents prepared by CONSULTANT pursuant to this Agreement are instruments of service, which shall be deemed the property of the CITY. CITY acknowledges and agrees that all plans, specifications, reports and other design documents prepared by CONSULTANT pursuant to this Agreement shall be used exclusively on this Project and shall not be used for any other work without the written consent of CONSULTANT. In the event CITY and CONSULTANT permit the reuse or other use of the plans, specifications, reports or other design documents, CITY shall require the party using them to indemnify and hold harmless CITY and CONSULTANT regarding such reuse or other use, and CITY shall require the party using them to eliminate any and all references to CONSULTANT from the plans, specifications, reports and other design documents. If a document is prepared by CONSULTANT on a computer, CONSULTANT shall prepare such document in a Microsoft® Word 97 SR-2 or lower format; in addition, CONSULTANT shall provide CITY with said document both in a printed format and on a three and one-half inch (3 1/2") floppy diskette.

ARTICLE 6
GENERAL PROVISIONS

6.1 Representation

A CITY representative shall be designated by the City Manager and a CONSULTANT representative shall be designated by CONSULTANT as the primary contact person for each party regarding performance of this Agreement.

6.2 Fair Employment Practices/Equal Opportunity Acts

In the performance of this Agreement, CONSULTANT shall comply with all applicable provisions of the California Fair Employment Practices Act (California Government Code Sections 12940-48) and the applicable equal employment provisions of the Civil Rights Act of 1964 (42 U.S.C. 200e-217), and the Americans with Disabilities Act of 1992 (42 U.S.C. § 11200, et seq.).

6.3 Personnel

CONSULTANT represents that it has, or shall secure at its own expense, all personnel required to perform CONSULTANT's services under this Agreement. Any person who performs engineering services pursuant to this Agreement shall be licensed as a Civil Engineer by the State of California and in good standing. CONSULTANT shall make reasonable efforts to maintain the continuity of CONSULTANT's staff who are assigned to perform the services hereunder and shall obtain the approval of the Director of Public Works of all proposed staff members who will perform such services. CONSULTANT may associate with or employ associates or subconsultants in the performance of its services under this Agreement, but at all times shall be responsible for their services.

6.4 Conflicts of Interest

CONSULTANT agrees not to accept any employment or representation during the term of this Agreement or within twelve (12) months after completion of the work under this Agreement which is or may likely make CONSULTANT "financially interested" (as provided in California Government Code Section 1090 and 87100) in any decisions made by CITY on any matter in connection with which CONSULTANT has been retained pursuant to this Agreement.

6.5 Legal Action

(a) Should either party to this Agreement bring legal action against the other, the case shall be brought in a court of competent jurisdiction in Los Angeles County, California, and the party prevailing in such action shall be entitled to recover its costs of litigation, including reasonable attorneys' fee which shall be fixed by the judge hearing the case and such fee shall be included in the judgment.

(b) Should any legal action about the Project between CITY and a party other than CONSULTANT require the testimony of CONSULTANT when there is no allegation that CONSULTANT was negligent, CITY shall compensate CONSULTANT for its testimony and preparation to testify at the hourly rates in effect at the time of such testimony.

6.6 Assignment

This Agreement shall not be assignable by either party without the prior written consent of the other party.

Notwithstanding the above, CONSULTANT may use the services of persons and entities not in CONSULTANT'S direct employ, when it is appropriate and customary to do so. Such persons and entities include, but are not necessarily limited to, surveyors, specialized consultants, and testing laboratories. CONSULTANT'S use of subcontractors for additional services shall not be unreasonably restricted by the CITY provided CONSULTANT notifies the CITY in advance.

6.7 Independent Contractor

CONSULTANT is and shall at all times remain, as to the CITY, a wholly independent CONTRACTOR. Neither the CITY nor any of its agents shall have control over the conduct of CONSULTANT or any of the CONSULTANT's employees, except as herein set forth. CONSULTANT expressly warrants not to, at any time or in any manner, represent that it, or any of its agents, servants or employees, are in any manner agents, servants or employees of CITY, it being distinctly understood that CONSULTANT is, and shall at all times remain to CITY, a wholly independent contractor and CONSULTANT's obligations to CITY are solely such as are prescribed by this Agreement.

6.8 Hazardous Materials

Unless otherwise provided in this Agreement, CONSULTANT and its subconsultants and/or contractors shall have no responsibility for the discovery, presence, handling, removal or disposal of, or exposure of persons to hazardous materials in any form at the site of the Project.

6.9 Titles

The titles used in this Agreement are for general reference only and are not part of the Agreement.

6.10 <u>Extent of Agreement</u>

This Agreement represents the entire and integrated Agreement between CITY and CONSULTANT and supersedes all prior negotiations, representations or agreements, either written or oral. This Agreement may be modified or amended only by a subsequent written agreement signed by both parties.

6.11 <u>Notices</u>

All notices pertaining to this Agreement shall be in writing and addressed as follows:
If to CONSULTANT:

      Name of the company
      Address

If to CITY:

      Mr. Ray Holland, Interim Director of Public Works
      City of Rancho Palos Verdes
      30940 Hawthorne Blvd.
      Rancho Palos Verdes, CA 90275

IN WITNESS WHEREOF, the parties hereto have executed this Agreement as of the date and year first above written.

Dated: _____   _____

                                             BY: _____

                                             Title: _____

Dated: _____   CITY OF RANCHO PALOS VERDES
A Municipal Corporation

                                             BY: _____

                                             MAYOR City of Rancho Palos Verdes

ATTEST:

_____

CITY CLERK

EXHIBIT "B"

CITY OF RANCHO PALOS VERDES

PUBLIC WORKS DEPARTMENT

JANUARY 2006

REQUEST FOR PROPOSAL

FOR

ENGINEERING DESIGN SERVICES

The City of Rancho Palos Verdes requests proposals from selected engineering firms for services related to the Engineering Design of Storm Drainage Systems along Palos Verdes Drive East in the area of Via Colinita and Roan Road.

2006 CITY OF RANCHO PALOS VERDES WATER QUALITY AND FLOOD PROTECTION PROGRAM DESIGN PROJECT

PROJECT SPECIFIC INFORMATION

The firm that is selected by the City will be expected to perform the following tasks, termed collectively as "design", for this project:

1. Communicate regularly with City staff regarding project issues via telephone, electronic mail, fax, and post mail.
2. Attend meetings: The Consultant will be required to attend, as a minimum, a design kickoff meeting, 50% and 95% design review meeting, and meetings with City as may be needed in support of any required environmental planning and/or permit application requirements. The City may require additional meetings, as needed, to resolve unforeseen issues or to discuss problematic design obstacles that arise.
3. Perform necessary site reconnaissance and investigations and review existing reports and as-built documents to determine existing conditions and design parameters.
4. Produce and provide to the City a video record of the site conditions at the start of design. Video shall be provided in digital video (*.mpeg or similar approved) format.
5. Perform necessary field surveying to determine existing grades, locations of design control points, and locations of existing features that may affect the design.
6. Review existing data, perform engineering analysis, make recommendations, design and provide plans and specifications for four projects to include: the complete removal of existing piping and installation

of new piping in various sizes at three locations in the Via Colinita Road area of the City; and for the removal of approximately 1,250 feet of 48 inch RCP and restoration of a natural channel in the Via Colinita Road and Miraleste Drive area of the City; and for the complete removal of existing piping and installation of new piping of varying sizes in the Roan Road area of the City; and for the complete removal of existing piping and installation of new piping in the Miraleste and Chandeleur Drive are of the City. Consultants designs should include recommendations regarding any needed improvements to inlets and outfalls; each of the plans and specification packages must include all required construction activities including demolition of existing facilities installation of new piping, inlets, catch basins, outlet structures, energy dissipaters, natural channel restoration requirements – improvements to safely (i.e. minimizing future erosion) discharge of the runoff and stabilization and restoration of lands disturbed by this construction.

7. Identify any additional easements that will be required for the construction and future maintenance of the designed facilities. Prepare legal descriptions and surveyor's sketches for the identified easements.

8. Coordinate with an Environmental Consultant as required in the preparation of California and Federal Environmental documentation. Incorporate requirements to protect and/or restorative efforts into the project plans and/or specifications.

9. Prepare and provide one reproducible bid ready set of all bid documents including construction-drawings (plans on mylar), specification package (using the City's standard format) and engineer's estimate. Provide City with an electronic copy of all bid documents; drawings in Autocad, specifications in Word, Estimates in Excel. NOTE: The City can provide standard contract and bidding form templates for consultant's use.

10. Respond to questions from prospective bidders and issue addenda, as needed.

FORM OF PROPOSAL

The Proposal shall be prepared in a "two sealed envelope" format.

The first sealed envelope shall contain the technical proposal and shall include:

1. Qualifications and experience of the key individual assigned to this contract.
2. Qualifications and experience of the firm.
3. Project Approach using the outline of required work as listed above limited to a maximum of five (5) pages.

The second envelope shall contain the cost proposal and include:

1. Fixed fee proposal for the work, subtotaled by the outline of required work as listed above.
2. The names of the Consultant's personnel who are to be involved, their titles, and hourly rates for their services.
3. Hourly Fees for all members of the firm.

EVALUATION PROCESS

Technical proposals will be evaluated on the following criteria:
1. Experience of key individual assigned to the contract.
2. Experience of firm in performing this type of service.
3. Understanding of project as demonstrated by the thoroughness of the proposal, the introduction of innovative or cost-saving ideas and approach.

4. References from clients for whom similar work in size and scope was performed.
5. Depth of staff available to perform services.

Any questions regarding the Request for Proposal shall be in writing to City of Rancho Palos Verdes Dept of Public Works, 30940 Hawthorne Blvd., Rancho Palos Verdes, CA 90275, Attention: Ron Dragoo, Senior Engineer/Project Manager.

Please note that the responses to this Request for Proposal are subject to the following conditions:

Insurance
The City of Rancho Palos Verdes requires the selected firm to have a minimum of $2,000,000 of professional errors and omissions insurance and $2,000,000 of general liability insurance prior to entering into an agreement with the City.

Acceptance of Terms
Submission of a proposal shall constitute acknowledgment and acceptance of all terms and conditions hereinafter set forth in the RFP unless otherwise expressly stated in the proposal.

Right of Rejection by the City
Not withstanding any other provisions of this RFP, the City reserves the right to reject any and all proposals and to waive any informality in a proposal.

Financial Responsibility
The proposer understands and agrees that the City shall have no financial responsibility for any costs incurred by the proposer in responding to this RFP.

Award of Contract
The selected firm shall be required to enter into a written contract with the City of Rancho Palos Verdes, in a form similar to the attached. This RFP and the proposal, or any part thereof, may be incorporated into and made a part of the final contract; however, the City reserves the right to further negotiate the terms and conditions of the contract with the selected proposer.

Exhibit A

Sample Contract

Exhibit B

Tentative Schedule including Submittals

Questions regarding this proposal shall be submitted in writing by 4:30 pm on February 21, 2006.

Proposals must be submitted by 4:30 pm on March 2, 2006.

Service Order General Conditions to Service Order Contracts

EXHIBIT A

SERVICE ORDER GENERAL CONDITIONS
TO SERVICE ORDER CONTRACTS

Part 1

The conditions stated in this Part I apply to all Service Order Contracts:

1.01 TIME OF ESSENCE, CHOICE OF LAW, TRADE MEANINGS, OWNERSHIP OF THE WORK

a. All time limits stated in the Contract Documents are of the essence of the Contract. Contractor shall commence Work immediately upon the execution of the Service Order Contract form and shall make every reasonable effort to complete its performance within the time or schedule provided for in the Scope or as may be approved by Owner.

b. The law of the state where the Job Site is located shall govern the interpretation and enforcement of this Contract.

c. Work described in words which so applied have a well-known technical or trade meaning shall be held to refer to such recognized standards.

d. All results arising from the Work are works made for hire for Owner within the meaning of the United States Copyright Act, and Contractor assigns to Owner the copyright and all other right, title and interest in the United States and elsewhere to all results arising from the Work. Contractor agrees to sign at any time at Owner's request any document necessary or useful to evidence or acknowledge such results as works made for hire for Owner and the assignment to Owner of all rights to such results.

1.02 INSTRUCTIONS

a. Contractor shall examine all Contract Documents and, if there is any ambiguity, shall ask Owner for clarification. All Work shall be executed in conformity with Owner's drawings, specifications, instructions and to the extent applicable all recommendations of the National Fire Protection Association regarding prevention of accidents, fires or for the elimination of any unsafe practice. Contractor shall do no Work without proper instructions.

b. Contractor shall comply with all relevant laws and regulations, federal, state and local as they apply to Contractor's performance of the Work. Without limiting the generality of the foregoing, Owner specifically directs Contractor to strictly comply with all laws and regulations relating to protection of the environment and public and employee health and safety. Except as otherwise expressly exempted in the Contract Documents, Contractor is strictly responsible to Owner for compliance with all such laws and regulations not only as they relate to the performance of the Work but also, in the case of worker safety and health rules, as they apply to the finished work and its suitability and capability to conform with applicable legal requirements for its intended use or operation by Owner. No provision in the Contract Documents shall be deemed to exempt Contractor from its obligation to comply with any particular law or regulation unless Owner expressly assumes the obligation to satisfy the specific compliance responsibility and such provision specifies the particular law or regulation, for which the compliance responsibility has been assumed by Owner.

c. Owner has adopted environmental requirements and procedures designed to reduce waste and resource utilization and maximize reuse and recycling opportunities. On-site Contractors are required to comply with Owner's environmental requirements. Each on-site Contractor shall review the environmental requirements for impact on the Work. The requirements documents may be secured from Owner's facility Environmental, Safety and Health ("ESH") Manager. Each off-site Contractor is encouraged to review the environmental requirements and advise Owner's respective facility or corporate ESH Manager of it's plan to conform to the applicable environmental requirements.

d. Contractor shall establish a Contractor Safety Program for the Work. To the extent applicable to the Work, and without limiting the generality of Contractor's obligations under paragraph 1.02(b) above, Contractor shall comply with the OSHA Process Safety Management Standard and the OSHA Hazard Communication Program Standard as specified in Section 2.16 below. Contractor is responsible for assuring that all of its subcontractors establish and implement, insofar as applicable to their respective work functions, appropriate safety programs and practices consistent with OSHA requirements. Owner may require Contractor to demonstrate that it has complied with all applicable safety requirements at such reasonable times as Owner may determine appropriate. Contractor shall be responsible for enforcing safety rules and shall maintain at all times, as required by the conditions and progress of the Work, all necessary safeguards for the protection of life, the Work, equipment on the Job Site, Owner's property and adjacent property and to assure uninterrupted production, and safe working conditions. Contractor shall require it's employees to comply with owner's substance abuse policy as well as any other safety practices Owner has established by plant rule or otherwise deems appropriate or necessary in specific circumstances. Contractor and/or any Subcontractor failing to follow safety rules shall be subject to eviction from the Job Site, may be refused re-entry, and may be terminated in accord with these General Conditions.

e. Contractor shall at all times enforce strict discipline among its employees, shall cause its employees to conform to all Owner Regulations, and shall not employ any unfit person or anyone not skilled in the Work assigned to it. Contractor shall not permit its employees, nor any of its Subcontractors to permit their employees to consume or possess on the Job Site any alcoholic beverage.

f. Owner may enforce badge identification at the Job Site.

1.03 MATERIALS AND PERFORMANCE

Unless otherwise stipulated, Contractor shall provide and initially pay for all Work. All goods supplied as part of the Work are guaranteed to be new, and the Work shall be of good quality, free from defects or fault and shall be prepared in a workmanlike manner.

1.04 PATENT LIABILITY CLAUSE

Contractor agrees to defend any claim, action or suit that may be brought against Owner, or its officers, agents or employees for infringement of any Letters Patent of the United States arising out of the performance of this Contract or out of the use or disposal by or for the account of Owner of supplies furnished or Work performed hereunder, and also to indemnify and hold harmless Owner and its officers, agents, and employees against all judgments, decrees, damages, costs and expenses recovered against it or them or sustained by it or them on account of any such actual or alleged infringement.

1.05 CHANGES

Owner, without invalidating the Contract, may order changes in the Scope which may result in an addition to or deduction from the compensation to be paid to Contractor. Owner may also order minor changes in the Scope, which in no specific, concrete or substantial way increase or decrease the Work; and such minor changes shall not involve any adjustment in compensation. All changes shall be evidenced by an executed Modification Order referencing the original Service Order Contract. If adjustments to compensation are required, they shall be stated in the Modification Order. The Contract may not be amended or modified except by Modification Order executed by both parties. Except as expressly amended by Modification Order, the Contract shall continue in full force and effect.

1.06 ADJUSTMENTS TO COMPENSATION

a. In the event of a change in Scope which requires additions to or deductions from Contractor's compensation, the parties agree to determine the adjustment, at Owner's option, in one or more of the following ways:

1. By estimate and acceptance of a lump sum amount.
2. By unit prices named in the Contract Documents or subsequently agreed upon,
3. By cost (time and material) plus percentage overhead and fee or
4. If agreed upon, a reimbursable basis subject to a "Not-to-Exceed" or a guaranteed maximum amount.

b. Percentage overhead and fee on such adjustments for Contractor and its Subcontractors if required for pricing methods 3 or 4 above shall be calculated as follows:

1. Overhead on the change in straight time labor cost shall be ten percent (10%) of the amount of such change.
2. Fee on the change in straight time labor cost shall be eight percent (8°%) of the sum of the amount of such change and the amount of the overhead calculated in the foregoing subparagraph.
3. Combined overhead and fee on the change in purchased material or leased equipment cost shall be ten percent (10%) of the amount of such change.
4. Combined overhead and fee on the change in the cost for subcontracted Work shall be five percent (5%) of the amount of such change.

c. In order to arrive at the value for any change, Contractor shall credit Owner with its projected cost(s), overhead and fee for any Work which was previously included, but which has been excluded by any such change.

d. Once the basis for an adjustment in compensation has been directed by Owner or agreed to by the parties a Modification Order memorializing such adjustment shall be signed by the parties. The procedures set forth in this Section are designed to encourage timely settlement and avoid litigation of claims concerning whether instructions by the Owner during the Work justify a change in the Contractors compensation, the schedule, or both. The procedures in this Section are also intended to attribute a final effect to the settlement of any claim between Owner and Contractor. This change procedure is mandatory and a condition precedent to any action Contractor shall file, if the issues in the action could have been addressed through the procedures set forth in this Section. If Contractor commences litigation concerning the value or effect of a directive or order of Owner without following the procedures established in this Section, or if Contractor institutes litigation concerning the value of a change previously agreed to by Contractor through a Modification Order signed by Contractor, then Contractor shall be liable to Owner for all attorney's fees and expenses Owner incurs in such litigation, regardless of the outcome of the litigation.

1.07 PREMIUM LABOR

All premium labor costs are to be borne by Contractor unless Owner has, in writing, directed Contractor to incur premium labor. In such event, Owner shall pay Contractor only that portion of the cost of the premium labor which is in excess of the actual rate for weekday Work plus applicable taxes or other charges which are based upon payroll amount. No increase will be allowed for Contractor's overhead and fee. Such premium time must have prior approval by Owner.

1.08 TAXES, INVOICING, PAYMENT AND AUDIT RIGHTS Except as separately provided in the Terms of Compensation:

a. Contractor shall pay any sales or use tax which may be determined to be applicable to the goods, materials and services to be provided by Contractor hereunder. Contractor shall contact the "Buyer" identified on the face of the Service Order form to obtain any required tax exemption registration number.

b. On or about the first day of each month, Contractor, shall submit to Owner at the address indicated on the face of the Service Order form, in duplicate, invoices for the Work completed the previous calendar month, less the amount to be retained by Owner which shall be ten percent (10%) of the value of the Work completed. Upon completion of the Work, Contractor shall submit in duplicate, a final invoice for the unpaid balance, including retainage withheld. Each invoice must show the total amount of Work complete to date, the amounts previously invoiced, and the payment requested. Except as otherwise provided herein, Owner shall make payment within thirty (30) days of invoice receipt.

c. For a period of at least two years after the completion of the Work, Contractor shall maintain such records as are necessary to substantiate that all invoices and applications for payment hereunder were valid and properly chargeable to Owner. Supporting documentation may include but is not necessarily limited to: time sheets, payroll records, receipts, contracts, books of account, banking records, and all such other records as reasonably relate to the Work. Owner or its representatives shall, upon reasonable prior notice to Contractor, be given the opportunity to audit such records in order to verify the accuracy of such invoices and other charges. Contractor shall maintain all such records in accordance with generally accepted accounting practices.

d. Contractor hereby represents and warrants that no payments have been or shall be made, directly or indirectly, by or on behalf of Contractor to or for the benefit of any employee or agent of Owner who may reasonably be expected to influence the decision to requisition, issue or take any action with respect to this Contract. Owner shall have the right to examine, at Owner's expense, such of Contractor's books and records as may be necessary, in the auditor's opinion, to verify Contractor's compliance with this paragraph. (As used herein, "payments" shall include money, property services and all other forms of consideration.) Owner's audit rights under this paragraph shall survive two years after the completion of the Work.

1.09 PAYMENTS WITHHELD

Owner may withhold or on account of subsequently discovered evidence nullify the whole or a part of any invoice to such extent as may be necessary to protect Owner from loss on account of any failure of Contractor to perform the Work in accordance with the Contract Documents, or satisfy obligations to its Subcontractors.

1.10 DELAYS AND EXTENSION OF TIME

If Contractor be delayed at any time in the progress of the Work by an act or omission of Owner or of any third party employed by Owner or by strikes, lockouts, fire, unusual delay in transportation, Acts of God or other causes over which Contractor has no control, then the time of completion shall be extended for such reasonable time as the parties shall decide, and no adjustment shall be made in Contractor's compensation. No such extension shall be made for delay occurring more than seven (7) days before claim therefore is made in writing to Owner. In the case of a continuing cause of delay, only one claim is necessary.

1.11 OWNER'S RIGHT TO TERMINATE CONTRACT

a. If Contractor persistently or repeatedly refuses or fails to conform to the provisions of the Contract Documents, then Owner may, without prejudice to any other right or remedy, terminate this Contract, take possession of the Job Site, and all supplies, materials and equipment thereon and finish the Work as Owner may deem expedient. In such case, Contractor shall not be entitled to receive any further payment until the Work is finished. If the expense of finishing the Work, including compensation for additional managerial and administrative services, is less than any compensation then remaining due to Contractor for Work performed prior to the date of termination, then such difference shall be paid to Contractor. If such expense exceeds such unpaid compensation, Contractor shall pay the difference to Owner. In the event suit is filed by Owner against Contractor to enforce provisions of the Contract Documents or to recover any loss resulting from Contractor's failure to perform, and Owner is successful, Contractor shall be liable, in addition to such other damages to which Owner may be entitled, for all attorney fees, court costs, and other expenses incurred by Owner in prosecution of the suit.

b. Owner reserves the right in its sole discretion and without cause or default on the part of Contractor, to terminate the Contract by providing written notice of such termination to Contractor. Upon receipt of such notice from Owner, Contractor shall immediately cease all Work. In the event of any such termination without cause, Owner shall compensate Contractor for all Work performed or obligations incurred through the date of termination at Contractor's cost, plus a reasonable fee on Work actually performed. In no event shall Contractor have a claim for loss of bargain damages.

1.12 ASSIGNMENT AND SUBCONTRACTS

a. This Contract shall be binding upon Owner and Contractor, and their respective successors and assigns. Contractor shall not assign this
Contract nor subcontract any Work hereunder without the prior written consent of Owner. Contractor is responsible for negotiating its contracts with Subcontractors and shall, to the extent applicable to the Work to be done by each Subcontractor, bind each by the terms of the Contract Documents.

b. Contractor shall not make nor commit to make any advance or prepayment to any Subcontractor without, in each instance, first obtaining Owner's specific prior written consent. Any losses due to defalcations by Subcontractors in connection with any advance or prepayment made in breach of the foregoing provision shall be incurred at Contractor's sole expense.

1.13 INDEMNITY

a. Definitions

For purposes of this Section 1.13
1. "THE OWNER" means and includes Owner, all of its subsidiaries and their respective affiliates, servants and employees;

2. "THE CONTRACTOR" means and includes Contractor and its Subcontractors and their respective servants, agents and employees; and

3. "LOSS" means any and all loss, damage, liability, or expense, whether incurred as a judgment, settlement, penalty, fine or otherwise (including attorney's fees and the cost of defense), in connection with any action, proceeding or claim, whether real or spurious, for injury, including death to any person or persons or damage to, loss of the use of, or loss of the property of any person, firm or corporation including the parties hereto, arising or resulting out of the performance of services required pursuant to this Contract.

b. The Contractor hereby agrees to indemnify and hold harmless THE OWNER from any and all LOSS to the extent caused or incurred as a result of the negligence or other actionable fault of THE CONTRACTOR.

c. Waiver of Certain Defenses With respect to the Owner's indemnity rights under the Contract Documents, Contractor expressly waives all statutory or common law defenses including but not limited to those under Worker's Compensation, Contribution, Comparative Fault or similar statutes or legal principles to the extent said defenses are inconsistent with or would defeat the purpose of the indemnifications provided under this Section 1.13 and to the extent applicable, Section 2.18.

d. Consideration It is stipulated and agreed that 1% of the Contractor's compensation is paid as and for specific consideration for this indemnity.

1.14 CASUALTY INSURANCE

a. Contractor hereby agrees to furnish and maintain at its own cost and expense, insurance policies underwritten by good and solvent insurance companies, protecting:

1. The legal liability of Contractor under the Worker's Compensation Act of any State or other statute or law, to pay claims for personal injuries sustained by its employees, including death resulting there from. Employer's Liability coverage shall be included with a limit of not less than $1,000,000.00.

2. The legal liability of Contractor to pay claims for damage to property, and for injuries to or death of any person or persons (including automobile exposure) and including the contractual liability assumed by Contractor under Section 1.13, and to the extent applicable Section 2.18, said insurance to be written with a limit of liability of not less than $2,000,000.00 for any one occurrence.

b. Owner shall be included under Contractor's insurance, required under a.2 above, as an additional insured with respect to claims and/or liability arising out of Work performed for Owner by Contractor or acts or omissions of Owner in connection with its general supervision of Contractor's Work. All insurance provided to Owner by Contractor shall be primary and any insurance maintained by Owner shall be excess and not contributing with Contractor's insurance.

c. Contractor shall not violate, or permit to be violated, any conditions of any of said policies, and shall at all times satisfy the requirements of the insurance companies writing said policies and shall furnish Owner with insurance certificates indicating the coverage referred to in subparagraph a. and b. above. Each such certificate shall state that Owner will be given thirty (30) days advance notice of cancellation or of any material restriction of coverage. Contractor shall be responsible for replacing cancelled coverage so that no hiatus in coverage occurs.

d. Contractor shall obtain either occurrence or claims made coverage. If Contractor has obtained claims made coverage to satisfy the foregoing insurance requirements, then Contractor agrees to maintain such claims made coverage for three years beyond the performance of this order.

1.15 WARRANTIES
Contractor represents and warrants that:

a. All information supplied by Contractor under this Contract is or shall be in accordance with Contractor's best information and belief, true and correct as of the date when given;

b. Except to the extent that Section 2.08 of these conditions applies to the Work, all Work performed under this Contract shall be free from defects for 90 days from the date of final payment by Owner. If within such time any defect is found, at Owner's request, Contractor shall correct same at Contractor's expense; and

c. Contractor has full legal right and authority to enter into and perform this Contract.

1.16 ADDITIONAL OBLIGATIONS
Owner and Contractor each agree to cooperate with the other in every respect in achieving the Scope. Specifically but without limitation, Contractor covenants with Owner to furnish his best-skill, care and judgment and at least the prevailing standard of skill, care and judgment expected of any similar Contractor under similar circumstances. Contractor agrees to furnish efficient business administration and superintendence and to use every effort to keep upon the Work at all times an adequate supply of workmen and materials, and to secure execution of the Work in the best, soundest, most expeditious and economical manner consistent with the interests of Owner.

1.17 CONTRACTOR'S STATUS
Contractor's relationship to Owner shall be that of an independent contractor. Neither party shall be deemed to be nor shall either represent itself to be the agent of the other.

1.18 NOTICES
Any written notice required or permitted under the terms of the Contract Documents (except as specifically provided for in the case of Insurance) shall be given and deemed to have been duly served when posted in prepaid U.S. registered mail, addressed to the respective party as set out on the Service Order Contract form.

1.19 NO WAIVER BY OWNER
The failure of Owner in any one or more instances to insist upon strict performance of any of the terms of the Contract Documents or to exercise any rights conferred therein, shall not be construed as a waiver or relinquishment to any extent of the right to assert or rely upon any such terms or rights on any future occasion.

1.20 APPROVALS
Wherever in the Contract Documents the consent or approval of either party is required before the other may take any particular action, such consent or approval must be obtained in writing and each party agrees not to unreasonably withhold same.

1.21 RIGHT TO INSPECT PERSONAL PROPERTY
No personal property other than clothes, lunch boxes, tool boxes, and similar items of a personal nature may be brought on to or removed from the Job Site without proper authorization. All property and vehicles being removed from the Job Site shall be subject to inspection.

1.22 CONFIDENTIALITY

Contractor hereby covenants that:

a. Contractor (i) shall maintain, as Owner's confidential property, (ii) shall not use for Contractor's own benefit, and (iii) shall not disclose, in whole or in part, to any third party (other than under subparagraph d below) any information, materials, or experience regarding Owner, its operation, its personnel, or its activities, including, but not limited to general or specific aspects of the Work provided hereunder (the "Information").

b. The duty of secrecy in subparagraph a shall not apply to Information:
  1. in the public domain through no fault of Contractor;
  2. obtained by Contractor in writing from a third party lawfully in possession of it and under no secrecy obligation as to it; or
  3. which was in Contractor's knowledge and possession prior to commencement of the Work (a) as a matter of right, (b) without restriction as to use or disclosure, and (c) not acquired by Contractor, either directly or indirectly, from Owner.

For purposes of this subparagraph b, no Information shall be deemed in the "public domain," or in Contractor's "knowledge and possession" merely because such Information is embraced by more general information. No right or license under any patent or copyright is being granted under this Agreement.

c. All tangible forms of Information, including, without limitation, all summaries, copies and excerpts of any Information, whether prepared by Contractor or not, shall be the sole property of Owner, and shall be immediately delivered by Contractor to Owner upon completion or termination of the Contract or upon Owner's request, whichever first occurs.

d. Contractor shall restrict access to the Information to only individuals who have a need-to-know in order to perform the Work, and Contractor shall require such individuals to agree, in writing, in form substantially similar to the provisions of this Section 1.22, to maintain all information so received in confidence.

e. Contractor's confidentiality obligations hereunder also shall extend to Information acquired from Owner in any and all invitations to bid on Work or agreements to provide services whether received or entered into prior to or after the date of this Contract.

f. Contractor's confidentiality obligations hereunder shall survive completion or termination of this Contract.

g. Contractor recognizes and acknowledges that any breach of this confidentiality provision would injure Owner irreparably, the amount of damage being impossible to ascertain. Owner may, therefore, in addition to pursuing any and all remedies provided by law, obtain an injunction against Contractor from any court having jurisdiction, restraining any violation of this provision.

1.23 ENTIRE AGREEMENT

The Contract Documents constitute the whole of the Contract between the parties hereto and neither has been induced to make or enter into this Contract by reason of any promise, agreement, representation, statement or warranty other than as contained in the Contract Documents.

1.24 DISPUTES

The parties agree that before, and as a condition precedent to, the initiation of any legal action or proceeding, all claims, controversies and disputes ("Disputes") arising out of or in relation to the performance, interpretation, application or enforcement of this Contract, including, without limitation, any breach hereof, the following process must be completed. In the event of a Dispute, the parties agree that their respective counsel and their project representatives familiar with the issue including Contractor's Chief Operating officer will schedule a meeting (by telephone or in person) to discuss the Dispute and to attempt in good faith to resolve it. If the matter is not resolved at such meeting, the parties may with in the next 60 days agree to private mediation of the dispute. If the Dispute is not then submitted to mediation, or not resolved after the completion of such Mediation, the parties may then pursue any available legal remedy, action, or proceeding.

1.25 JURY TRIAL WAIVER

Contractor and Owner hereby mutually and expressly waive their rights to a jury trial in connection with any litigation involving a Dispute solely between them arising out of this Contract.

Excerpts from a Typical Construction Contract

ARTICLE 2 – PERFORMANCE OF THE WORK

2.01   PERFORMANCE OF WORK - GENERAL

Contractor shall, at its own cost and expense, furnish all necessary materials, labor, transportation, and equipment for doing and performing said work and the materials used shall comply with the requirements of the Contract documents. In addition, the Contractor shall perform, with its own organization, Contract work amounting to at least fifty percent (50%) of the Total Bid Amount. All work shall be performed and completed as required in the Contract documents, and subject to the approval of the Engineer, or his/her designated assistant.

2.02   NO ASSIGNMENT OR DELEGATION

Contractor shall not assign or delegate the duties or obligation under this Contract or his/her interest therein in whole or in part without the prior written consent of the City, which may be withheld at the City's sole discretion.

2.03   STANDARD OF PERFORMANCE

Contractor agrees that all services performed hereunder shall be provided in a manner commensurate with the highest professional standards and shall be performed by qualified and experienced personnel; that any work performed by Contractor under the Contract will be performed in the best manner; that any material furnished shall be subject to the approval of the Engineer; and that both work and materials will meet fully the requirements of these Plans and Specifications.

2.04   DEFECTIVE WORK

The Contractor shall remove and rebuild at his/her own expense any part of the work that has been improperly executed, even though it has been included in the monthly estimates. If he/she refuses or neglects to replace such defective work, prior to acceptance of the work, it may be replaced by the City at the expense of the Contractor, plus 15% for overhead expenses, and his/her sureties shall be liable therefore. (See Section 2.14 for curing defects after acceptance of the work.)

2.05   COMMUNICATIONS REGARDING THE WORK

After award of the Contract, all communications regarding the work covered by these Specifications shall be addressed to the City Engineer, City of Glendale, and mailed or hand delivered to:

> City of Glendale
> City Engineer's Office
> 633 E. Broadway, Room 205
> Glendale, California 91206

2.06   INDEPENDENT CONTRACTOR

The Contractor in the performance of the work hereunder will be acting in an independent capacity and not as an agent, employee, partner, or joint venturer of the City.

Editorial Note: Next section(s) deleted from original contract for educational purposes.

2.13   GUARANTEE OF WORK

A. The Contractor guarantees all materials and workmanship against defects for a period of one year, unless noted otherwise, from the date of final acceptance of all work performed under the Contract. The date of final acceptance will be established and confirmed in writing by the Engineer.

B. The Contractor assumes responsibility for a one-year guarantee, unless noted otherwise, for all work and materials provided or performed by subcontractors, manufacturers, or suppliers.

C. The Contractor hereby agrees that if, within a period of one year, unless noted otherwise, after final acceptance of the work done under the Contract, any portion of the work installed, constructed, or performed fails to fulfill any of the requirements of the Contract, he/she will, without delay and with the least practicable inconvenience and without further cost to the City, repair or replace defective or otherwise unsatisfactory work or materials. This Contract will not delay acceptance of the work or final payment.

D. Should the Contractor fail to act promptly in accordance with this requirement, or should the exigencies of the case require repairs or replacements to be made before the Contractor can be notified or can respond to notification, the City may at its option make the necessary repairs or replacements, or perform the necessary work, and the Contractor shall pay to the City the actual cost of such repairs plus 15 percent, or City may charge such costs to the Contractor's retention.

E. The Contractor shall be responsible for the full expense incidental to making good any and all of the above guarantees and Contracts. The above guarantees and Contracts are covenants, the performance of which shall be binding upon the Contractor and his/her sureties.

Editorial Note: Next section(s) deleted from original contract for educational purposes.

ARTICLE 3 – TIME OF COMMENCEMENT AND COMPLETION

COMMENCEMENT, PROSECUTION, AND COMPLETION OF WORK

A. NOTICE TO PROCEED

The Contractor is not authorized to perform any work under these Specifications until he/she has received from the City an official notification to commence work. The date on which the Contractor receives the notification is herein referred to as the Notice to Proceed. The Contractor shall commence work within 14 calendar days after Notice to Proceed. A copy of the Notice to Proceed is included in Exhibit C3.

The notification to commence wok will not be issued until the Contract is property executed, bonds are furnished and approved, and insurance has been submitted and approved.

B. PROSECUTION OF THE WORK

Work shall be continued at all times with such force and equipment as will be sufficient to complete it within the specified time. The Contractor expressly proposes that he/she has taken into consideration and made allowances for all ordinary delays and hindrances to the work to be performed and that he/she will complete the work within the specified time.

C. REQUIRED CONTRACT COMPLETION

The work must be completed in its entirety and made ready for service within Forty (40) working days following issuance of the Notice to Proceed.

3.02   CITY'S DISCRETION TO EXTEND TIME

In the event the work required hereunder is not satisfactorily completed in all parts and in compliance with the Contract Documents, the City shall have the sole right, in its discretion, to increase the number of working days or not, as may seem best to serve the interest of the City. A change order extending the Contract time only will be issued by the City should the City decide to increase the number of working days.

3.03   DELAYS AND EXTENSIONS OF TIME FOR CONTRACTOR

A. The Contractor shall take reasonable precautions to foresee and prevent delays to the work. In the event of any delay to the work, the Contractor shall revise his/her sequence of operations, to the extent possible under the terms of the Contract, to offset the delay.

B. If any delay to the work is caused by circumstances within the Contractor's control, it is not excusable and not compensable, and the Contractor will not be entitled to any extension of time or to any other compensation for damages resulting directly or indirectly therefrom.

C. If any delay having a direct effect on the work is caused by circumstances beyond the control of the Contractor except for causes of delay specified in Paragraph 3.03-D., such delay may be excusable and may entitle the Contractor to an equivalent extension of time, but not to any other compensation. Excusable but not compensable causes include but are not limited to labor disputes, weather conditions unfavorable for prosecution of the work, and force majeure.

D. If any delay having a direct effect on the work is caused by failure of the City to provide information as specified, or necessary instructions for carrying on the work, or to provide the necessary right of way or site for installation, or failure of a utility to remove or relocate an existing facility such delay may be compensable and may entitle the Contractor to an equivalent extension of time, and to compensation for damages resulting directly from any of the causes of delay specified in this paragraph.

Editorial Note: Next section(s) deleted from original contract for educational purposes.

3.07   LIQUIDATED DAMAGES

A. The deduction for liquidated damages shall be $500.00 per working day from date of required Contract completion until actual Contract completion date.

B. The above liquidated damages are necessary to ensure timely completion and to defray costs of additional construction and contract administration. Timely completion is required to insure that the owner may occupy the building fully, all facilities operational and all construction activities completed in accordance with these Specifications.

C. Should the Contractor fail to complete all or portions of the contract work within the specified completion date, liquidated damages in the amount of mentioned above shall be assessed until such work is complete and accepted by the City. Such assessments shall be deducted from Contractor's funds being held in the Contractor's account or provided by the Contractor if the Contractor's account does not have sufficient funds.

D. It being impracticable or extremely difficult to fix the actual damage, the amount set forth above is hereby agreed upon as liquidated damages and will be deducted from any money due the Contractor under this Contract. Should the amount of the damages exceed the amount due the Contractor, he/she and his/her sureties shall be liable for the excess.

10.01   PROTECTION OF PERSONS AND PROPERTY

Contractor's Responsibility: Notwithstanding any other provision of these Specifications, the Contractor shall be solely and completely responsible for conditions of the job site, including safety of all persons and property, during performance of the work. This requirement will apply continuously and will not be limited to normal working hours. Safety and sanitary provisions shall conform to all applicable Federal, State, County, and local laws, regulations, ordinances, standards, and codes. Where any of these are in conflict, the more stringent requirement shall be followed.

10.02   PROTECTION FROM HAZARDS

A. **Trench Excavation**

   Excavation for any trench four (4) feet or more in depth shall not begin until the Contractor has received approval from the Engineer of the Con tractor's detailed plan for worker protection from the hazards of caving ground during the excavation of such trench. Such plan shall show the details of the design of shoring, bracing, sloping or other provisions to be made for worker protection during such excavation. No such plan shall allow the use of shoring, sloping or protective system less effective than that required by the Construction Safety Orders of the Division of Occupational Safety and Health, and if such plan varies from the shoring system standards established by the Construction Safety Orders, the plan shall be prepared and signed by an engineer who is registered as a Civil or Structural Engineer in the State of California.

B. **Confined Spaces**

Contractor shall comply with all of the provisions of General Industry Safety Orders of the California Code of Regulations. Entry of a confined space shall not be allowed until the Contractor has received approval from the Engineer of the Contractor's program for confined space entry. Confined space means a space that (1) is large enough and so configured that an employee can bodily enter and perform assigned work; and (2) Has limited or restricted means for entry or exit (for example, tanks, vessels, silos, storage bins, hoppers, vaults, and pits are spaces that may have limited means of entry); and (3) is not designed for continuous employee occupancy. Failure to submit a confined space entry program may result in actions as provided in Article 5; "SUSPENSION OR TERMINATION OF CONTRACT", of these Specifications.

C. **Material Safety Data Sheet**

Contractor shall comply with all of the provisions of General Industry Safety Orders of the California Administrative Code. The Contractor shall submit to the Engineer a Material Safety Data Sheet (MSDS) for each hazardous substance proposed to be used, ten (10) days prior to the delivery of such materials to the job site or use of such materials at a manufacturing plant where the Engineer is to perform an inspection. For materials which are to be tested in City laboratories, the MSDS shall be submitted with the sample(s). Hazardous substance is defined as any substance included in the list (Director's List) of hazardous substances prepared by the Director, California Department of Industrial Relations, pursuant to Labor Code Section 6382. Failure to submit an MSDS for any hazardous substance may result in actions as provided in Article 5, "SUSPENSION OR TERMINATION OF CONTRACT," of these Specifications.

Editorial Note: Next section(s) deleted from original contract for educational purposes.

ARTICLE 11 – INDEMNITY, INSURANCE AND BONDS

11.01 INDEMNITY

To the fullest extent permitted by law, Contractor hereby agrees, at its sole cost and expense, to defend, protect, indemnify, and hold harmless the City of Glendale ("City"), and its officers, employees, elected officials, attorneys, members of boards and commissions, agents, and volunteers (hereinafter collectively referred to as "Indemnitees") from and against any and all damages, costs, expenses, liabilities, claims, demands, causes of action, proceedings, expenses, judgments, penalties, liens, and losses of any nature whatsoever, including fees of accountants, attorneys, or other professionals and all costs associated therewith (hereinafter collectively referred to as "Liabilities"), arising or claimed to arise, directly or indirectly, out of, in connection with, resulting from, or related to any act, failure to act, error, or omission of Contractor or any of its officers, agents, servants, employees, subcontractors, suppliers or their officers, agents, servants or employees, arising or claimed to arise, directly or indirectly, out of, in connection with, resulting from, or related to the Contract or the performance or failure to perform any term, provision, covenant, or condition of the Contract, including this indemnity provision. This indemnity provision is elective regardless of any prior, concurrent, or subsequent active or passive negligence by Indemnitees, except that, to the limited extent mandated by California Civil

Code Section 2782, the Contractor shall not be responsible for liabilities which arise from the sole negligence or willful misconduct of Indemnitees or arise from the active negligence of City. This indemnity provision shall survive the termination of the Contract or final payment hereunder and is in addition to any other rights or remedies which City may have under the law or under the Contract. Payment is not required as a condition precedent to an indemnitee's right to recover under this indemnity provision, and an entry of judgment against an Indemnitee shall be conclusive in favor of the Indemnitee's right to recover under this indemnity provision.

11.02   PUBLIC LIABILITY AND PROPERTY DAMAGE INSURANCE

Contractor shall assume all responsibility for damages to property or injuries to persons, including accidental death, which may arise out of or may be caused by Contractor's performance of this Contract, by its subcontractor, or by anyone it directly or indirectly employed, and whether the damage or injury may accrue, or may be discovered, before or after termination of this Contract.

Contractor shall obtain, pay for, and maintain during the life of this Contract an "occurrence" policy for Commercial General Liability, including Contractual Liability, and an Automobile Liability policy which shall protect it and the City from claims for injuries and damages. The insurance company shall be an "admitted" insurer in the State of California; shall be domiciled within, and organized under the laws of, a State of the United States; and shall carry an A.M. Best & Company minimum rating of "A:VII." The policy shall name the City of Glendale, its officers, agents, and employees as additional insureds under the policy in the following amounts:

1. COMMERCIAL GENERAL LIABILITY INSURANCE in the amount not less than TWO MILLION DOLLARS ($2,000,000.00) per occurrence for personal injuries, including accidental death, to any one person; property damage insurance in an amount not less than TWO MILLION DOLLARS ($2,000,000.00) and subject to the above limits and combined single limit of insurance in an amount not less than TWO MILLION DOLLARS ($2.000,000.00); and

2. AUTOMOBILE LIABILITY INSURANCE endorsed for all vehicles, whether hired, scheduled, owned or non-owned, with a combined single limit of at least TWO MILLION DOLLARS ($2,000,000.00) per occurrence for personal injuries, including accidental death, to any one person; and for property damage.

All policies shall contain a "Severability of Interest" clause and a "Primary Coverage" clause for any loss arising out of or caused by Contractor's performance of the Contract. In addition, the policies shall contain a statement of obligation on the carrier's part to notify the City, by registered mail, at least 30 days in advance of any policy cancellation, termination, or reduction in the amount of coverage. Contractor shall furnish the City with a "certificate of insurance" and an "additional insured endorsement," both documents countersigned by the insurance carrier or its authorized representative, on forms satisfactory to the City Attorney, which set forth the above provisions.

The countersigned certificate, along with the additional insured endorsement ... officers, agents, and employees are named as additional insureds under this policy." This insurance is primarily to the coverage of the City of Glendale. Neither the city nor any of its insurers shall be required to contribute to any loss. This policy contains a severability of interest clause. The issuing company shall mail 30 days advance notice to the City of any policy cancellation, termination, or reduction in the amount of coverage.

The certificate of insurance must also state if any claim has been paid or is currently pending under the policy, and if so, the amount of such claim(s). Any deductibles or self-insured retentions shall be set forth on the certificate and shall be subject to the City's review and approval.

<u>CONTRACTOR SHALL FURNISH THE REQUIRED CERTIFICATE(S) OF INSURANCE AND ENDORSEMENTS(S) BEFORE THE CITY SIGNS THIS CONTRACT.</u>

11.03   CONTRACTOR'S FAILURE TO OBTAIN, PAY FOR, OR MAINTAIN INSURANCE

Contractor's failure to obtain, pay for, or maintain any required insurance shall constitute a material breach upon which the City may immediately terminate or suspend this Contract. At its discretion, the City may obtain or renew the insurance and pay all or part of the premiums. Upon demand, Contractor shall repay the City all monies the City has paid to obtain or renew the insurance, or the City may offset the cost of the premium against any monies due Contractor from the City.

11.04   WORKER'S COMPENSATION INSURANCE

Contractor understands and agrees that all persons furnishing services to the City under this Contract are, for the purpose of workers' compensation liability, employees solely of Contractor and not of the City. Contractor shall bear the sole responsibility, liability, and costs of furnishing workers' compensation benefits to its employees, or anyone it directly or indirectly employs, for injuries arising out of or connected with services performed on behalf of Contractor.

Contractor shall obtain, pay for, and maintain for the duration of this Contract, complete workers' compensation insurance. The City of Glendale shall not be responsible for any claims at law or in equity caused by Contractor's failure to comply with this paragraph. Contractor shall provide proof of workers' compensation coverage by furnishing the City with an insurance certificate or a certificate of consent to self-insure.

All workers' compensation insurance policies shall bear an endorsement, or shall attach a rider, which provides that the carrier will notify the City, by registered mail, at least 30 days in advance of the policy's expiration, termination, or cancellation.

11.05   BUILDERS RISK/COURSE OF CONSTRUCTION INSURANCE

A. The Contractor shall obtain and maintain during the life of the Contract, in the name of the Contractor and the City, "all risk" builders risk/course of construction insurance issued on a completed value form covering the entire work of the Contract for 100% of the replacement value thereof without deductibles. Such insurance shall include items of labor and materials in place or to be used as part of the permanent construction, surplus miscellaneous materials and supplies incident to the work, temporary buildings, debris removal, and false work, staging, towers, forms and equipment, whether or not owned or rented by the Contractor, and demolition required by the enforcement of any Contract, law or regulation. Such insurance shall also include coverage for damage due to earthquake and flood.

B. In the event of a loss by the perils insured against, of any or all of the work and/or materials herein provided for, at any time prior to the final completion of the Contract and the final acceptance by the City, the Contractor shall promptly reconstruct, repair, replace or restore all work or materials so destroyed.

C. Nothing herein provided for shall in any way excuse the Contractor or his/her surety from the obligation of furnishing all the required materials and completing the work in full compliance with the terms of the Contract.

The following endorsements/coverages shall be included in the policy:

1. The coverages are primary to any other applicable insurance coverages carried by the City.

2. The contractor and the insurance company waive all rights of subrogation and contribution against the City, its officers, employees, elected official, attorneys, members of boards and commissions, agents and volunteers; and

3. The policy stall not be canceled nor the above coverages/ endorsements reduced without 60 days written notice to the City.

11.06   ENDORSEMENTS AND CERTIFICATE

A. Contractor shall have its insurance carrier(s) complete and execute the following documents, together with a copy of each insurance policy required under the Contract, including all endorsements thereto, which shall be delivered to the City Engineer within 14 working days following issuance of the Notice of Award:

1. Form entitled: Commercial General Liability/Automobile Liability/Special Endorsement, attached hereto as Exhibit "A1,"

2. Proof of Workers' Compensation Insurance, or if Contractor is self-insured for worker's compensation, a self-insuring certificate therefor from the State of California.

3. Form entitled: Certificate of Insurance, with 30 days prior notice of cancellation required as to all coverage and policies.

B. The Contract will not be executed by the City and the Notice to Proceed issued until the aforesaid insurance documents have been received and approved by the City. City's decision as to the acceptability of all insurance documents is final. No substitution of the form of the documents or the endorsements or amendments thereto will be permitted without the prior written consent of City.

11.07   BONDS

A. Contractor shall furnish the following bonds:

1. A Faithful Performance Bond in an amount equal to One Hundred percent (100%) of the Contract price in the form shown in Exhibit "B1" attached hereto.
2. A Payment Bond (Labor and Material) in an amount equal to One Hundred percent (100%) of the total Contract price in the form shown in Exhibit "B2" attached hereto.

3. A Maintenance Bond in an amount equal to Ten percent (10%) of the total Contract price in the form shown in Exhibit "B3" attached hereto.

B. All such bonds shall be accompanied by a power of attorney from the surety company authorizing the person executing the bond to sign on behalf of the company. If the bonds are executed outside the State of California, a California representative of the surety must countersign all copies of the bonds. The signature of the person executing the bond shall be acknowledged by a Notary Public as the signature of the person designated in the power of attorney.

Editorial Note: Next section(s) deleted from original contract for educational purposes.

Master Agreement Between Inland Architects, Inc. and Big Corp, Inc. for Architectural/Consulting Services Alliance

Table of Contents

ARTICLE I. PREAMBLE

THIS MASTER AGREEMENT ("Agreement") is between Inland Architects, Inc., a California Corporation ("Architect") and Big Corp, Inc., a Delaware Corporation ("BCI") each of which may be referred to in the singular as "Party" or in the plural as "Parties:" Subject to the terms and conditions of this Agreement, Architect and BCI agree that Architect will provide to BCI, an alliance of architectural services ("Services") in accordance with the terms and conditions stated herein.

Architect's Federal Taxpayer Identification Number is 99-1111111 and Architect's State License Number is C-9898989.

The Parties, intending to be legally bound, agree as follows:

1. Term of Agreement

This Agreement is effective on the date when signed by the last Party ("Effective Date"), and shall continue in full force and effect for a term of three (3) years unless sooner Terminated or Canceled as provided in this Agreement The Parties may extend the term of this Agreement by mutual agreement in writing.

The Termination, Cancellation or expiration of this Agreement shall not affect the obligations of either Party to the other Party pursuant to any Contract previously executed hereunder, and the terms and conditions of this Agreement shall continue to apply to such Contract as if this Agreement were still in effect.

2. Alliance

BCI desires to establish an Alliance relationship with the Architect. This relationship will be structured to draw on the strengths of each organization and to identify and achieve common goals for this Alliance. The Alliance objectives are:
- To build trust by fostering cooperation and openness through shared knowledge and understanding;
- To add value by providing Solutions that will lower total costs for both Parties;
- To establish performance measures to ensure continuous improvement; and
- To provide an ongoing resource to manage their relationship.

The Parties agree that they will use best efforts to work together to develop Key Performance Measures ("KPMs") that may measure such aspects of Architect's performance such as:
- Quality and cost of Services such as Change Orders and construction unit costs;
- Adherence to schedule;
- Billing errors;
- Client/Customer service;
- Compliance to contract requirements; and
- Cost reduction

A complete set of KPMs is to be developed by the Alliance Council and provided to the Parties.

ARTICLE II. SCOPE OF SERVICES

1. Access to Premises

   A. BCI's Premises

   For the purposes of Architect's performance of the Services hereunder, Architect and Architect's personnel shall have reasonable access to any BCI site where Services are to be performed, subject to rights of access secured or owned by BCI. Architect shall be responsible for Architect's personnel observing BCI's site rules and regulations and those of its property owners and property managers, including but not limited to, safety regulations and security requirements, and for working in harmony with others white present at the site. If BCI, for any reasonable and lawful reason, requests Architect to discontinue furnishing any person provided by Architect for performing Services on BCI's Premises, Architect shall immediately comply with such request. Such person shall leave BCI's Premises promptly, and Architect shall not furnish such person again to perform Services on BCI's Premises without BCI's consent.

   B. Architect's Premises

   BCI shall, upon reasonable prior notice to Architect and at no additional charge, have reasonable access to Architect's premises during normal business hours in order to observe Architect's work with respect to the Services hereunder.

   C. Releases Void

   Neither Party shall waive or release any personal rights from representatives or customers of the other in connection with visits to its Premises, unless required by a third Party property owner or property manager, and both Parties agree that no such release or waiver shall be presented as evidence by them or third persons in any legal action or proceeding.

2. General

   A. Architect's evaluations of BCI's Project budget and Architect's opinions of probable Construction Costs as provided for herein shall be made on the basis of Architect's experience and qualifications, and will represent Architect's professional judgment as a qualified design professional familiar with the construction industry. Since Architect has no control over the cost of labor, materials, equipment, or services furnished by others, or the competitive bidding process, or future market conditions, Architect does not guarantee or represent that proposals, bids, negotiated prices,, or actual Construction Costs wilt not vary from Architect's opinions of probable Construction Costs prepared or agreed upon by the Architect.

   B. Architect shall furnish or cause to be furnished, from time to time, design and technical Services, and other necessary Services, including Basic Services, Revisions and Extra Services, and Pass Through Projects (all collectively called "Services") as set forth in this Agreement and/or any Contract, for such construction and alteration Projects as BCI may assign to Architect.

   C. Architect shall be responsible for design accuracy of existing visible and readily accessible field conditions and will be held accountable for any change resulting from errors in such design (hereinafter referred to as "Design Errors")

D. Nothing contained herein shall obligate BCI to assign any Project or Projects to Architect.

3. Basic Services

Unless otherwise noted in this Agreement and/or any Contract, Architect shall perform or cause to be performed the "Basic Services" listed below which shall consist of the five phases described as appropriate to each Project, and include normal architectural, structural, mechanical, and electrical Services.

A. Schematic Design Phase

Architect shall consult with BCI to ascertain Project requirements. If requested by BCI, the Architect shall provide Project programming as an extra service to the contract on a time and material basis. If requested by the Architect, BCI shall provide the services of their maintenance personnel or a contractor, to remove or expose existing construction items or provide existing conditions information, if available. In addition, Architect shall provide the following Services:

1. Prepare a maximum of three (3) initial design studies and schematic Plans illustrating the scale and the relationship of Project components to be approved by BCI in writing.

2. Provide preliminary cost estimate in a format required by BCI. As deemed necessary by BCI, BCI shall make available the services of the Alliance General Contractor to provide assistance with this effort.

3. Attend a maximum of four (4) meetings in California during this phase.

4. Provide meeting minutes to all Parties requested by BCI.

5. Develop space plan layout for block plan work stations for the Approved schematic design.

6. Participate with electrical and mechanical Consultants to perform building load calculations and life cycle cost analysis for all design alternatives based on BCI's growth forecast data (if available), BCI's UVAC standards, building assessment survey or maintenance data (if available), and using acceptable software e.g., Trace or DOEII for load and life cycle cost analysis.

7. Additionally, Architect shall provide a copy of the new load calculations evaluated by electrical Consultants for standby engines.

8. Evaluate Interior elevations of critical areas as required.

9. Provide input on their relative time frames for Project schedule as well as projections for plan check and permit processing.

10. Verify extent of existing conditions with special attention to evaluation of existing HVAC and electrical systems as much as possible given availability of as-built documents and field exploration.

11. Coordinate/integrate efforts with other Consultants to ensure the most efficient and cost effective design.

12. For building addition Projects, make a minimum of two (2) visits to the planning department and fire department to confirm all special permitting requirements or variances required.

B. Design Development Phase

Architect shall prepare from Approved schematic design studies and Plans, design development documents to fix and describe the size and character of the entire Project as to structural, mechanical, and electrical systems, materials and such other essentials as may be necessary. Architect shall:

1. Provide BCI with a cost-effective design that will result in a cost-effective construction which meets BCI's requirements.

2. Define scope of Services from previous alternatives for 1-IVAC, electrical and other building systems.

3. If requested by BCI, Architect shall advise BCI of the selected interior/color finishes, prior to going out to bid for a Project. Architect will provide two (2) color board selections with one revision, which are included in the scope of Basic Services.

4. Make a minimum of one (1) visit to the Project site unless Architect secures written approval from BCI to forego such visit for a particular Project. Such approval shall not release Architect from responsibility to visit all other Project sites under this Agreement.

5. Attend a minimum of two (2) meetings and make a minimum of two (2) revisions to Drawings during the design phase prior to BCI's written Approval to begin construction documents.

6. Provide meeting minutes during design phase to all Parties as BCI may request.

C. Construction Document Phase

1. After the design development documents are Approved, Architect shall prepare Working Drawings in the form of:
   a. AutoCAD created reproducible master,
   b. AutoCAD diskettes, and
   c. A hard copy

2. AutoCAD drawings must be in conformance with the BCI Corporate Real Estate CAD Standards. The purpose of these CAD standards is to provide Corporate Real Estate (CRE) personnel, Consultants and Contractors standard guidelines for the development of BCI's CAD drawings.

3. These Working Drawings shall set forth in detail the requirements for the construction of the Project. In addition, the Architect shall be responsible for:
   a. Distribution of documents as directed by BCI (all copies of Drawings are a reimbursable cost and are not included in the Basic Services).
   b. Preparation of all reasonable design changes requested by BCI to Approved design documents during the construction phase at no additional cost to BCI.
   c. Completion of all building department reviews and submittals including obtaining governmental plan check approval of all phases of Project design from all governmental agencies typically charged with the responsibility of approving such design. Architect shall make such changes as necessitated by governmental agencies subsequent to BCI's review and written approval. Architect will make every effort to defend BCI's viewpoint when position of costs need to be contended with the governmental building department. All governmental permit and plan check fees are paid by the Architect and reimbursed at cost by BCI; and

d.   Signage details are part of Basic Services and are provided at no additional cost to BCI.

D. Bidding Phase

Upon BCI request, Architect shall interpret Drawings and Specifications and issue clarification and addenda as may be necessary. Architect shall attend one (1) pre-bid Site meeting.

E.   Construction Phase

This phase shall commence after BCI has issued a construction Contract for the particular Project. Architect shall:

1.   Make a minimum of one (1) visit to the construction site in order that Architect and his Consultants shall be familiar with the progress and confirm that all critical elements of construction are in strict conformance with the Drawings and Specifications;

2.   Monitor and immediately inform BCI of any critical elements of construction that are not in strict conformance with the Drawings and Specifications;

3.   Attend periodic meetings at the site, as mutually agreed upon by both BCI and Architect;

4.   Review construction schedules and material lists and recommend acceptance, modification, or denial;

5.   Review and confirm that the Shop Drawings, Product Data, and Samples submitted by the Contractor are in strict conformance with the intent of the construction documents;

6.   Review, when requested by BCI, Contractor's applications for payment and recommend acceptance modification or denial;

7.   Interpret the Drawings and Specifications, upon request by BCI or Contractor;

8.   Prepare price requests, submit them to the Contractor, and evaluate the proposed changes when requested by BCI;

9.   Review Contractor's costs and time extension proposals connected with Changes, and submit written recommendations to BCI;

10. Continually monitor construction to determine and prepare a list for BCI of the deficiencies that may prevent the planned occupancy of certain areas of the Project;

11. Review all design implementation verification tests (e.g. air balance reports, switch gear testing) and report any deficiencies to BCI;

12. Make a final site visit to observe and review the Project. Within ten (10) days after the final site visit, Architect must prepare and furnish to Contractor a written list of any deficiencies (Punch List) that may prevent the acceptance of the Project by BCI. Completion or correction of said deficiencies shall be done within such time limit as may be established by BCI. Architect shall review the Punch list for completion or correction after such time by making a follow-up site visit to confirm that the Punch list work has been completed;

F. Closeout Phase

    1. Obtain and review for completeness, submittal brochures, operating instructions, manuals, warranties, permits and the like, required by the Specifications or by governmental authority, obtain all record Drawings "as built", update all original AutoCAD Drawings to reflect the as-built conditions and submit all such items, including revised AutoCAD files on CD-ROM, in addition to any engineering calculations to BCI within 30 days from receipt of same from the Contractor(s) and prior to submitting a final bill for Basic Services and notify BCI in writing that all required documents have been completed and submitted to BCI;

    2. Attend one (1) Project audit and client review meeting;

    3. Deliver final Project invoice to BCI;

G. Items not expected in Basic Services

    1. Color detailed renderings are not expected to be in the Basic Services;

    2. Out of the ordinary or special feasibility/planning studies are not expected to be in the Basic Services.

4.   Revisions and Extra Services

Revisions and Extra Services shall include, but not be limited to, the following:

A. Changes requested by BCI's Representative either prior to, or subsequent to Approved schematic design documents, design development documents or construction documents that result in substantial work by Architect and such work would not have been required if BCI's Representative had not made such a request.

B. Development of supplemental Drawings and Specifications required by BCI's Representative for the purpose of obtaining alternate bid amounts, and may include any deductive or additive alternate.

C. Studies and reports requested by BCI or governmental agencies and are not required as part of Basic Services. These may include, but are not limited to: substantial time spent preparing and filing applications and reports, including without limitation, Environmental Impact Reports, Environmental Impact Statements, and Energy Commission Standards Studies, including but not limited to Regional Coastal Commissions and Historic Districts.

D. Design work requested by BCI's Representative during the construction phase necessary to obtain cost proposals from the Contractor for proposed construction Changes, excluding Work required as a result of Architect Design Error.

E. Space Planning shall be considered a Revision and Extra Service when it is a standalone Project. If Space Planning Services are required as part of an overall Project being contracted under Basic Services the contracting Parties shall, prior to the start of work, determine if it is reasonable to include such Space Planning Services within the compensation for Basic Services. If such Space Planning Services are not included within the compensation for Basic Services, then such Space Planning Services shall be included within the compensation for Revision and Extra Services.

F. All Revisions and Extra Services must be authorized by the appropriate BCI Representative, in writing and in accordance with BCI's Schedule of Authorizations prior to the start of such Services. BCI will not pay for Revisions or Extra Services that are improperly authorized.

G. Protracted Planning department approvals shall be considered an extra service when the time spent in dealing with this issue exceeds 20% of the combined budget for schematic design and design development.

H. At the direction of the BCI project manager and when a project construction schedule requires the bidding of a Project to be broken into multiple phases, the incremental effort associated with this task shall be considered an extra service for which time spent shall be billed on a time and material basis.

5. Permits and Licenses

If requested by BCI's Representative, Architect shall assist BCI in obtaining all permits and licenses required by any properly constituted public authority for the performance of the Services. Coordinating with adjacent property owners for access to adjacent property shall not be considered part of Basic Services.

6. Services Provided by BCI

Unless otherwise set forth under this Agreement and/or any Contract, if any, BCI may perform or furnish as necessary to the Architect the following:

A. Necessary information as to Project requirements;

B. Land survey of the site, if available to BCI

C. Soils engineering reports, if available to BCI

D. Laboratory testing and inspection reports; and

E. Sample invitations to bid.

ARTICLE III. COST AND COMPENSATION

1. Basic Services

As set forth in any Contract issued hereunder, Architect shall be compensated for Basic Services, by the method described next in Section 2 entitled "Fixed Percentage" For non-Basic Service (e.g., Studies and Space Plan that involves furniture layout) Architect shall be compensated by the method described in Section 3 entitled "Multiple of Payroll". The compensation method for non-Basic Services shall be determined by BCI and will be so identified on the Contract.

2. Fixed Percentage

A. Under this method, Architect shall initially be paid a certain percentage of the amount of the estimated Construction Costs for construction of the Project and, Architect's final fee shall be based on the Final Construction Cost of the Project less any Construction Cost resulting from Architect and/or Consultant Errors and Omissions. Such percentage shall be as set forth in ARTICLE VII. Exhibit

B, Matrix A and shall be paid monthly in proportion to Services performed so that the compensation at the completion of each phase shall equal the following percentages of the total compensation for Basic Services:

- Schematic Design Phase   10%
- Design Development Phase   20%
- Construction Document Phase   45%
- Bidding Phase   5%
- Construction Phase   15%
- Closeout Phase   5%
- Total Payments   100%

B. Billing through and including the construction document phase shall be based on the estimated Construction Cost stated on the Contract as "Construction Budget" or any subsequent estimated Construction Costs Approved by BCI's Representative.

C. Pass-through projects are compensated as a fixed percentage as set forth in ARTICLE VII. Exhibit B, Matrix C.

3. Multiple of Payroll

A. Under this option, Architect shall be reimbursed based on a multiple of employees' basic payroll costs. Titles, basic hourly rates and payroll multipliers for normal and overtime work by employees are as set forth in ARTICLE VII. Exhibit B. Basic hourly rates shall not include any loading or fringe benefits whether mandatory or voluntary. Total hourly reimbursement for employees' work, after application of the payroll multiplier, shall not exceed the Principal's hourly rate.

B. Architect shall be reimbursed for Services performed by contract workers based on a multiple of Actual Costs incurred as set forth in ARTICLE VII Exhibit B.

C. Architect shall be reimbursed for the Actual costs of Consultants such as architects, and structural, electrical, and mechanical Consultants retained by the Architect. Such Actual Cost shall be computed on the same basis as Architect's Actual Costs, as set forth in this Agreement including ARTICLE VII Exhibit B and shall not, without BCI's Representative's prior written approval, exceed the limits placed on Architect.

D. Architect shall be compensated for time spent by Principals as set forth in ARTICLE VII Exhibit B.

E. Contracts payable under this option may be subject to a "not to exceed" dollar amount. Such "not to exceed" dollar amount shall be so stated on the Contract.

4. Fee Schedule

Payment for Architect's Services shall be based upon the Schedule of Fees and Labor Rates described in ARTICLE VII Exhibit B.

If Architect determines that there are going to be cost overruns, Architect shall immediately notify BCI so that BCI and Architect may jointly identify the factors that affect the cost, and make necessary adjustments to control the cost collectively.

5. Revisions and Extra Services

There will be no additional reimbursement to Architect for any accepted additive alternate supplemental Drawings and Specifications for which the Basic Services are compensated. Architect shall be compensated for Revisions and Extra Services, which are not included in Basic Services, requested by BCI on the same basis as described in ARTICLE III Section 2 entitled "Fixed Percentage", except:

A. Architect's compensation for any Studies and reports shall be a mutually agreed upon stipulated sum.

B. In the event that Services are rendered for a Project, or a portion of a Project that is not pursued to completion by BCI, Architect shall be compensated on a "Multiple of Payroll" method or on a mutually agreed upon basis.

6. Reimbursable Expenses

Architect shall be reimbursed at Actual Cost as set forth in ARTICLE VII. Exhibit B-l.

7. Summary of Actual Costs

Architect shall provide BCI with a summary of Architect's Actual Costs annually and Architect shall make all supporting documentation for the summary available to BCI consistent with Article II, Section 7. The summary shall include the following data as a minimum.

A. Job titles, basic hourly rates, and payroll multipliers for normal overtime work by employees as set forth in the ARTICLE VII. Exhibit B. Basic hourly rates shall not include any loading or fringe benefits whether mandatory or voluntary. Total hourly rate for employee's work after application of the payroll multiplier shall not exceed the Principal's hourly rate.

B. Job titles, basic hourly rates, and payroll multipliers for work performed by Contract Worker as set forth in ARTICLE VII. Exhibit B.

C. Actual Costs incurred by the Architect for Services provided by structural, electrical, and mechanical engineers that were retained by the Architect.

D. Hours spent by the Principals as set forth in ARTICLE VII. Exhibit B.

ARTICLE IV. BILLING AND INVOICING

1. Invoicing

Subject to any provisions to the contrary, Architect shall render a summarized invoice for Services rendered under the applicable Contract to BCI, that comply with the following requirements:

A. Except as otherwise provided for in this section, invoices shall be submitted on the 5th and 19th calendar days of each month. If the 5th or 19th falls on a weekend or BCI Holiday, the invoices may be submitted on the next business day.

B. Architect shall submit a final true-up Change Order for a Project to BCI no later than thirty calendar days following receipt of final Project documents from Contractor.

C. Architect's final invoice for a Project (hereinafter referred to as "Final Project Invoice") shall be delivered to BCI no later than thirty calendar days following BCI approval of the Architect's true-up Change Order.

D. BCI shall have no obligation to pay Architect's Final Project Invoice not meeting requirements of this section (B and C above) unless failure to perform is due to causes beyond Architect's control

E. The invoicing shall be in a format and detail substantially as described in ARTICLE VII Exhibit D.

F. BCI shall pay Architect within forty-five (45) days of the date of receipt of the invoice in accordance with the prices set forth in this Agreement or in the applicable Contract.

G. The Architect shall not be entitled to any other compensation of any kind whatsoever unless such compensation is specifically approved in advance and in writing by BCI.

H. BCI may deduct any setoff or recoupment claims that it or its Affiliates may have against Architect from amounts due or to become due to Architect, whether under this Agreement or otherwise. Architect shall pay any amount due to BCI or its Affiliates that is not applied against the invoiced amounts within thirty (30) days after written demand by BCI.

2. Invoice Supporting Documentation

A. General

As indicated in *ARTICLE II. Section 7 "Records and Audits"*, Architect agrees to maintain supporting documentation for all charges billed to BCI hereunder and provide such documentation to BCI upon request. BCI shall not have any obligation to pay any invoice amount(s) that cannot be verified through supporting documentation.

For all invoices, Architect shall itemize all costs as specific categories requested by BCI's Representative. This includes reporting costs to:
   1. Special Studies
   2. Space Planning
   3. Basic Services
   4. Additional elements as may be requested, including but not limited to division by phase of job, floor, etc.,
   5. Reimbursable expenses

Invoicing under the provisions of ARTICLE III shall include when practicable, original receipts for monies expended by Architect and original invoices for monies billed to Architect. When impracticable, the duplicate or counterpart, or a reproduced copy endorsed "Original Bill Not Available" shall be accepted.

Supporting documentation shall also list names, categories and hourly rates of persons engaged in providing the Services (including Consultants if performing Services), identify the hours and phases of the Services (e.g., schematic design), total charges, and any applicable California sales or use taxes which shall be separately stated and identified.

B. Claims

Any claims for extension of time or for an increase in the Services' amount shall be made in writing to BCI not more than five (5) days after the commencement of the delay or the occurrence giving rise to the claim for extra compensation. In the case of a continuing delay, only one claim is necessary. The Architect shall provide an estimate of the probable effect of such delay on the progress of the Services.

C. Disputes Regarding Invoicing

If BCI disputes any invoice rendered or amount paid, BCI will so notify Architect and the Parties will use their best efforts to resolve such disputes expeditiously. Provided that if BCI so notifies Architect of a disputed invoice, then the time for paying the invoice in a dispute shall be extended for a period of time equal to the time between Architect's receipt of such notice from BCI and the resolution of such dispute or, if BCI has already paid the amount in dispute, BCI may notify Architect and withhold the amount in dispute from any subsequent payment until the resolution of such dispute.

ARTICLE V. PERFORMANCE

1. General

Architect shall be measured on the following performance categories during the term of this Agreement:

A. Design costs;

B. Errors and Omissions;

C. Construction Unit Costs;

D. Adherence to schedule;

E. Customer/Client Service; and

F. Design/Build Coordination.

The specific measurements are outlined in the KPMs.

2. Acceptance or Rejection

Any Services performed by the Architect pursuant to this Agreement shall be accepted by BCI unless BCI determines in good faith and with reasonable exercise of judgment, that such Services have not been performed in accordance with this Agreement and/or any Contract. BCI shall notify Architect of any instance in which Services are not so performed. Upon receipt of notice, Architect shall take prompt and appropriate corrective action. Architect shall not charge BCI for any actions taken to correct Services not performed in accordance with this Agreement and/or any Contract. If unsatisfactory Services have not been corrected within a reasonable time, not to exceed ten (10) working days after BCI's notice to Architect, or within a time period agreed to in writing by the Parties, or if Architect's Services are performed unsatisfactorily two (2) or more times within any sixty (60) day period, BCI may, in addition to all other rights and remedies provided by law or this Agreement, Cancel this Agreement and/or the Contract.

As stated in ARTICLE I. Section 3 entitled "Alliance", the Architect's performance will be evaluated with the use of KPMs. The Architect shall demonstrate an ability to meet these KPMs over a stated period of time. Should the Architect fail to meet the minimum requirements of the KPMs, BCI shall have the right to cancel this Agreement upon (30) thirty calendar days prior written notice.

3. Termination, Cancellation and Suspension of Services

    A. Termination
    Either Architect or BCI may terminate for convenience this Agreement upon thirty (30) days prior written notice to the other setting forth the effective date of such termination; provided, however, that no such termination shall be effective prior to the expiration of the first twelve (12) months of this Agreement. The Termination, Cancellation or expiration of this Agreement shall not affect the obligations of either Party to the other Party pursuant to any Contract previously executed hereunder, and the terms and conditions of this Agreement shall continue to apply to such Contract as if this Agreement had not been terminated, canceled or expired.

    B. Cancellation
    If Architect is in material default of any of its obligations hereunder, including but not limited to, Architect being adjudged bankrupt, filing or having filed against Architect a petition in reorganization proceeding; or Architect making a general assignment for the benefit of creditors; or a receiver being appointed for Architect's business because of Architect's insolvency or inability to meet current obligations; or if, in the sole opinion of BCI, at any time after execution of this Agreement and/or Contract, Architect refuses or neglects, without fault of BCI, to provide professional services sufficient to complete the Services within the time specified in this Agreement and/or any Contract; or if, in the sole opinion of BCI, Architect fails in any material respect to conduct the Services with promptness, safety and diligence, or to perform any of the provisions hereof, and such default is not corrected or reasonably commenced to be corrected within three (3) days after written notice thereof; or any default occurs two times within a sixty (60) day period; BCI may, in addition to all other rights and remedies provided by law and this Agreement, 1) cancel this Agreement and/or any Contract, and/or 2) have the Services completed immediately by BCI employees, or by any other third Party at Architect's sole cost and expense.

    C. Cease Work
    Architect shall immediately cease performing any Services or incurring any costs under this Agreement and/or any Contract as of the effective date of such termination, cancellation or suspension of services.

    D. Consequences
    Architect shall be entitled to compensation for Services accepted by BCI and satisfactorily completed prior to the effective date of the termination, cancellation or suspension of Services. In no case shall such compensation exceed the amount, if any, agreed upon in this Agreement and/or any Contract for such Services. In the event of suspension of Services, Architect shall not be entitled to receive further payment hereunder until the Services shall be wholly finished. Architect agrees to waive and hereby does waive all claims against BCI for lost profits, losses, damages or other economic injury arising out of such termination, cancellation or suspension of Services. BCI's decision to terminate, cancel or suspend under this provision shall not be subject to mediation.

    E. Additional Consequences of Termination or Cancellation

If BCI terminates, cancels or suspends this Agreement and/or any Contract pursuant to the provisions of this Agreement, Architect shall waive all claims against BCI for termination, cancellation or suspension charges, or lost profits, losses or damages. Upon termination, cancellation or other expiration of this Agreement:

1. BCI and Architect shall release the other from any further obligation, duty or liability arising out of this Agreement and/or any Contract, as the case may be, except for any liability or obligation, whether of indemnity or otherwise, which may have accrued or which may be accruing at the time of the termination, cancellation, or other expiration, or which is otherwise specified in this Agreement and/or Contract.

2. Architect shall promptly return to BCI, all papers, written materials, properties, other materials and other information furnished to Architect by BCI or developed by Architect in connection with or as a result of performance of the Services under this Agreement and/or any Contract. Each Party shall provide the other such reasonable assistance as may be necessary for the orderly, non-disrupted business continuation of each Party.

F. Representations

1. In addition to any other rights or remedies BCI may have at law or in equity, the architect perform or cause to be performed the Services under this Agreement in accordance with usual and customary professional care and with generally accepted architectural/engineering practices in effect at the time the service are performed. The Services provided under this Agreement and/or any Contract issued hereunder shall be performed by properly qualified personnel in a diligent, workmanlike manner, to BCI's satisfaction in accordance with this Agreement, in a manner consistent with professional standards, and in compliance with applicable Laws and Regulations. Neither payments to Architect nor total or partial occupancy of the Project location by BCI or persons operating under Agreement with BCI shall relieve Architect of these obligations.

2. Notwithstanding anything to the contrary herein contained, it is understood and agreed that the foregoing representation shall not affect, limit, or impair BCI's right against Architect, or any other Party Architect causes to perform professional services under this Agreement, and their sureties, if any, with regard to latent defects in the Services that are not apparent, and that could not by the exercise of reasonable care and due diligence, be ascertained or discovered by BCI upon visual inspection.

4. Time is of The Essence

Architect understands that BCI's business and operations require the performance of Services hereunder be accomplished in the period of time and by dates specified in the applicable Contract. Therefore, it is agreed that TIME IS OF THE ESSENCE in the performance of Services. In determining a period of time, the date upon which action is taken to start the period shall not be counted and the period shall end with the close of business on the last designated day of the period. If Architect learns of anything that might prevent the timely performance of Services, Architect will immediately notify BCI of all relevant information concerning the potential delay

5. Reporting

A. At BCI's request. Architect shall deliver written and/or verbal Project status reports to BCI.

B. In the event KPMs requires information to be delivered by the Architect, Architect shall deliver such information to BCI in a timely manner.

ARTICLE VI. GENERAL PROVISIONS

1. Amendments and Waivers

This Agreement and any Contracts placed hereunder may be amended or modified only through a subsequent written document signed by the Parties; provided that BCI may, at any time, make changes to the scope of Work, and Architect shall not unreasonably withhold or condition its consent. An equitable adjustment shall be made if such change substantially affects the time of performance or the cost of the Work to be performed under this Agreement. Such cost adjustment shall be made on the basis of the actual cost of the Work, unless otherwise agreed in writing. No course of dealing or failure of either Party to strictly enforce any term, right or Condition of this Agreement shall be construed as a general waiver or relinquishment of such term, right or condition. A waiver by either Party of any default shall not be deemed a waiver of any other default.

2. Assignment

Neither Party hereto may assign, subcontract or otherwise transfer its rights or obligations under this Agreement except with the prior written consent of the other Party hereto, provided, however that BCI will have the right to assign this Agreement to any present or future Affiliate, subsidiary or parent corporation of BCI, without securing the consent of the Architect and may grant to any such assignee the same rights and privileges BCI enjoys hereunder. Any attempted assignment not assented to in the manner prescribed herein, except an assignment confined solely to money due or to become due, will be void. It is expressly agreed that any assignment of money will be void if (1) Architect fails to give BCI at least thirty (30) days prior written notice thereof, or (2) such assignment imposes or attempts to impose upon BCI additional costs or obligations in addition to the payment of such money or (3) denies, alters or attempts to alter any of BCI's rights.

3. Compliance with Laws

Architect and all persons furnished by Architect shall comply with all applicable Laws and Regulations of the Fair Labor Standards Act (FLSA), Occupational Safety and Health Act (OSHA), and all other applicable foreign, federal, state, county and local laws, ordinances, regulations and codes, including but not limited to, the procurement of required permits, certificates, approvals, inspections, and licenses when needed in Architect's performance of Services under this Agreement. Architect further agrees to comply with all applicable Executive and Federal regulations, as set forth in "Executive Orders and Associated Regulations," a copy of which is attached hereto as Exhibit A, and by reference made a part of this Agreement. Architect further agrees to indemnify BCI for any charges, penalties, fines, losses, damages and other costs, including attorney's fees, arising out of, or related to Architect's noncompliance with the law. It will be the responsibility of the Architect and/or Consultants to design facilities and revisions to facilities in accordance with the laws listed in this section.

4. Conflict of Interest

Architect represents and warrants that no officer, director, employee or agent of SRC has been or will be employed, retained or paid a fee, or otherwise has received or will receive, any personal compensation or consideration, by or from Architect or any of Architect's officers, directors, employees or agents in connection with the obtaining, arranging or negotiation of this Agreement or other documents entered into or executed in connection with this Agreement.

5. Construction and Interpretation

A. The language of this Agreement shall in all cases be construed simply, as a whole and in accordance with its fair meaning and not strictly for or against any Party. The Parties agree that this Agreement has been prepared jointly and has been the subject of arm's length and careful negotiation. Each Party has been given the opportunity to independently review this Agreement with legal counsel and other consultants, and each Party has the requisite experience and sophistication to understand, interpret and agree to the particular language of the provisions. Accordingly, in the event of an ambiguity in or dispute regarding the interpretation of this Agreement, the drafting of the language of this Agreement shall not be attributed to either Party.

    B. Article, section and paragraph headings contained in this Agreement are for reference purposes only and shall not affect the meaning or interpretation of this Agreement. The use of the word "include" shall mean "includes, but is not limited to." The singular use of words shall include the plural and vice versa. All obligations and rights of the Parties are subject to modification as the Parties may specifically provide in a Contract. "Services" and "Software" shall be treated as "goods" for purposes of applying the provisions of the Uniform Commercial Code ("UCC"). If there is an inconsistency or conflict between the terms in this Agreement and in a Contract, the terms in the Contract shall take precedence for that Contract only.

6. Dispute Resolution

    A. The Parties will attempt in good faith to promptly resolve any controversy or claim arising out of or relating to this Agreement through negotiations between authorized representatives of the Parties, before resorting to other remedies available to them.

    B. If a controversy or claim should arise which is not settled as specified in Subsection A., representatives of each Party who are authorized to resolve the controversy or claim will meet at a location designated by BCI, at least once, and will attempt to, and are empowered to resolve the matter. Either representative may request this meeting within fourteen (14) days of such request (the "first meeting").

    C. Unless the Parties otherwise agree, if the matter has not been resolved within twenty-one (21) days of the first meeting, the representatives shall refer the matter to more senior representatives, who shall have full authority to settle the dispute. Such senior representatives will meet for negotiations within fourteen (14) days of the end of the twenty-one (21) day period referred to above, at a site designated by BCI. Three (3) business days prior to this scheduled meeting, the Parties shall exchange memoranda stating the issue(s) in dispute and their positions, summarizing the negotiations which have taken place, and attaching relevant documents.

    D. If more than one (1) meeting is held between the senior representatives, the meeting shall be held in rotation at the offices of Architect and BCI.

    E. If the matter has not been resolved within thirty (30) days of the first meeting of the senior representatives (which period may be extended by mutual agreement), the Parties will attempt in good faith to resolve the controversy or claim in accordance with the American Arbitration Association's rules for Mediation of Business Disputes.

7. Emergency Support

In the event any natural or other emergency/disaster occurs whereby the Services provided pursuant to this Agreement are rendered inoperative, Architect agrees to coordinate alternative procedures with BCI so as to minimize disruption and restore the Services provided herein.

8. Entire Agreement

This Agreement and/or any applicable Contract, together with any exhibits, attachment, schedules, and/or other documents, attached hereto or referenced in this Agreement which are hereby incorporated by reference, constitute the entire Agreement between the Parties with respect to the subject matter contained therein. Notwithstanding the above, Architect's proposals attached or referenced herein, are hereby incorporated by reference, as part of this Agreement and/or any Contract solely to the extent that the proposal defines the scope of the Services, the Specifications in excess of those set forth in this Agreement and/or any Contract, and the Fee Schedule if applicable. Any and all other additional or different terms in Architect's proposal, quotation, acknowledgment, bill, forms, or other communication, whether or not such terms materially alter this Agreement, shall be deemed objected to by BCI without need of further notice of objection, and shall be of no effect and in no circumstance binding upon BCI unless such terms and conditions are expressly accepted by BCI in writing as an amendment to this Agreement and/or any Contract in accordance with ARTICLE VI. Section 2 entitled Amendments and Waivers."

9. Force Majeure

Neither Party shall be deemed in default of this Agreement or any Contract to the extent that any delay or failure in the performance of its obligations results from any cause beyond its reasonable control, and without its fault or negligence; such as acts of God, acts of civil or military authority, embargoes, epidemics, wars, riots, insurrections, fires, explosions, earthquakes, floods, unusually severe weather conditions or strikes. If any force majeure conditions occur, the Party delayed or unable to perform will give immediate notice thereof to the other Party. The Party affected by the other's inability to perform may elect to:

A. Terminate this Agreement and/or any Contract, in whole or in part, as to Services not already performed.

B. Suspend the affected Contract or any part thereof for the duration of the force majeure condition, with the option to obtain elsewhere the Services to be furnished under such Contract(s), and deduct from any commitment under such Contract(s) the quantity of Services obtained or for which commitments have been made elsewhere.

C. Resume performance hereunder once the force majeure condition ceases with an option in the affected Party to extend the period of this Agreement or any affected Contract up to the length of time the force majeure condition endured.

Unless written notice to the contrary is given within thirty (30) days after such affected Party is notified of the force majeure condition, option B. above will be deemed selected.

10. Governing Law

This agreement and performance hereunder shall be governed by the laws of the state of California exclusive of its choice of laws provisions.

If a dispute arises and the material facts affect contracts submitted by more than one company the governing law shall be California.

11. Indemnity

A. TO THE FULLEST EXTENT PERMITTED BY LAW, ARCHITECT SHALL DEFEND, INDEMNIFY AND HOLD: HARMLESS BCI AND ITS AFFILLATES (INCLUDING THEIR EMPLOYEES, OFFICERS, DIRECTORS, AGENTS AND CONTRACTORS) AGAINST ANY LIABILITY ARISING FROM OR INCIDENTAL TO ARCHITECT'S OBLIGATIONS UNDER THIS AGREEMENT OR THE MATERIAL OR SERVICES PROVIDED BY ARCHITECT, INCLUDING (i) INJURIES TO PERSONS, INCLUDING DEATH OR DISEASE, (ii) DAMAGES TO PROPERTY, INCLUDING THEFT, (iii) ARCHITECT'S FAILURE TO COMPLY WITH ALL LAWS, AND (iv) LIENS ON BCI'S PROPERTY.

B. IT IS THE INTENT OF THE PARTIES THAT THIS INDEMNITY APPLY REGARDLESS OF WHETHER OR NOT SUCH LIABILITY WAS CAUSED IN PART BY BCI'S OWN NEGLIGENCE OR THAT OF THE OTHER PARTIES INDEMNIFIED UNDER THIS SECTION, EXCLUDING ONLY ANY LIABILITY ARISING FROM THE SOLE NEGLIGENCE OF BCI. THIS INDEMNITY SHALL SURVIVE THE DELIVERY, INSPECTION AND ACCEPTANCE OF THE MATERIAL OR SERVICES.

C. BCI SHALL NOTIFY ARCHITECT WITHIN A REASONABLE PERIOD OF TIME OF ANY WRITTEN CLAIM, DEMAND, NOTICE OR LEGAL PROCEEDINGS ("CLAIM") FOR WHICH ARCHITECT MAY BE RESPONSIBLE UNDER THIS INDEMNITY OBLIGATION. A DELAY IN NOTICE SHALL NOT RELIEVE ARCHITECT OF ITS INDEMNITY OBLIGATION, EXCEPT TO THE EXTENT ARCHITECT CAN SHOW IT WAS PREJUDICED BY THE DELAY.

D. ARCHITECT SHALL ASSUME, AT ITS EXPENSE, THE SOLE DEFENSE OF THE CLAIM THROUGH COUNSEL SELECTED BY ARCHITECT AND SHALL KEEP BCI FULLY INFORMED AS TO THE PROGRESS OF SUCH DEFENSE; UPON REASONABLE REQUEST OF ARCHITECT AND AT ARCHITECT'S EXPENSE, BCI SHALL COOPERATE WITH ARCHITECT IN THE DEFENSE OF THE CLAIM. AT ITS OPTION AND EXPENSE, SEC MAY RETAIN OR USE SEPARATE COUNSEL TO REPRESENT IT, INCLUDING IN-HOUSE COUNSEL. ARCHITECT SHALL MAINTAIN CONTROL OF THE DEFENSE, EXCEPT THAT IF THE SETTLEMENT OF A CLAIM WOULD ADVERSELY AFFECT SEC. ARCHITECT MAY SETTLE THE CLAIM AS TO BCI ONLY WITH ITS CONSENT, WHICH CONSENT SHALL NOT BE WITHHELD OR DELAYED UNREASONABLY. ARCHITECT SHALL PAY THE FULL AMOUNT OF ANY JUDGMENT, AWARD OR SETTLEMENT WITH RESPECT TO THE CLAIM AND ALL OTHER EXPENSES RELATED TO THE RESOLUTION OF THE CLAIM, INCLUDING COSTS, INTEREST AND REASONABLE ATTORNEYS' FEES. IF BCI IS REQUIRED TO TAKE ANY ACTION TO ENFORCE ITS INDEMNITY RIGHTS UNDER THIS AGREEMENT, OR TO ASSUME THE DEFENSE OF ANY CLAIM FOR WHICH IT IS ENTITLED TO RECEIVE AN INDEMNITY UNDER THIS AGREEMENT, BECAUSE OF ARCHITECT'S FAILURE TO PROMPTLY ASSUME SUCH DEFENSE, THEN BCI MAY ALSO RECOVER FROM ARCHITECT ANY REASONABLE ATTORNEYS' FEES (INCLUDING COST OF IN-HOUSE COUNSEL AT MARKET RATES FOR ATTORNEYS OF SIMILAR EXPERIENCE) AND OTHER COSTS OF ENFORCING ITS INDEMNITY RIGHTS OR ASSUMING SUCH DEFENSE.

E. ARCHITECT AGREES NOT TO IMPLEAD OR BRING ANY ACTION AGAINST BCI OR BCI'S EMPLOYEES BASED ON ANY CLAIM BY ANY PERSON FOR PERSONAL INJURY OR DEATH THAT OCCURS IN THE COURSE OR SCOPE OF EMPLOYMENT OF SUCH PERSON BY ARCHITECT AND RELATES TO ARCHITECT'S PERFORMANCE UNDER THIS AGREEMENT.

12. Independent Contractor

Architect hereby represents and warrants to BCI that:

A. Architect is engaged in an independent business and will perform all obligations under this Agreement as an independent contractor and not as the agent or employee of BCI;

B. Architect's personnel performing Services shall be considered solely the employees of Architect and not the employees or agents of BCI;

C. Architect will be responsible for Architect's acts and those of Architect's personnel during the performance of Architect's obligations under this Agreement.

13. Information

A. In the performance of its obligations under this Agreement, Architect and its personnel may receive or have access to Specifications, original tracings, engineering calculations, proposals, research, records, reports, recommendations, Drawings, standards, sketches, manuals, findings, evaluations, models, forms, reviews, Samples, tools, computer programs, disks, diskettes, basic detailed manipulatable source information regarding BCI's buildings and facilities in any form including, without limitation, computer-aided design data and diskettes, technical information, data, confidential business, customer or personnel information or, written, oral or otherwise confidential information (all hereinafter referred to as "Information") owned or controlled by BCI.. Such Information, in whole or in part, may be subject to Laws and Regulations regarding secrecy of communications or trade secrets, or may be proprietary and/or confidential or may be disclosures of patentable inventions with respect to which patents may not have been issued or for which patent applications may not have been filed.

B. Accordingly, Architect agrees: that no part of the Information shall be reproduced in any form or by any electronic or mechanical means including information storage and retrieval systems without permission obtained in writing from BCI; to prevent reproduction of the Information; and to inform its personnel engaged in handling such Information of the proprietary or confidential character of such Information and of the existence of applicable Laws and Regulations regarding secrecy of communications. Unless such Information was previously known to Architect free of any obligation to keep it confidential or has been or is subsequently made public by BCI or a third Party, it shall be kept confidential by Architect, shall be used only in performing Services hereunder, and may be used for other purposes only upon such terms as may be agreed upon in writing.

C. Except with BCI's prior written consent, information that Architect may disclose hereunder to BCI shall be deemed non-confidential, non-proprietary, and free from all restrictions on use or disclosure. If Architect provides BCI with any proprietary or confidential information which is conspicuously marked as such, BCI shall use the same degree of care to prevent its disclosure to others as BCI uses with respect to its own proprietary or confidential information. Architect shall label or otherwise mark all of its confidential information in the following manner:

PROPRIETARY INFORMATION

The information contained herein is for use by authorized employees only and is not for general distribution with Big Corp, Inc.

D. Any Services performed in connection with this Agreement and/or any Contract is proprietary and confidential. Architect shall not disclose any aspect of the Services to any third person, business, or governmental agency without prior written approval of BCI, except as otherwise required by Laws and Regulations, and then only upon immediate written notification to BCI so that BCI may legally challenge such disclosure.

E. If the Services include the submission of a written report to BCI, such report will be deemed confidential. In that report, Architect shall include a description of the Services completed, a compilation of the data generated, results of the data, conclusions that may be derived from the results, recommendations for any additional Services that may be necessary including remedial programs, and additional information as requested by BCI. Prior to submittal of any final report, Architect shall prepare draft copies of the report which shall be presented in writing or, at BCI's request, verbally to:

> Big Corp, Inc.,
> 100 E. Smith Ave
> Interior CA 91000
> Attn: Area Manager, Los Angeles Service Area
> Fax: 626-310-8031

> OR:

> Big Corp, Inc.,
> 2600 South 5th Street
> San Pablo, CA 91000
> Attn: Contract Manager-Alliance Services
> Fax: 909-876-9801

Architect shall include two cover pages with each copy of any draft report. Architect shall type the word "Draft" on the first of the two cover pages. Architect shall meet with BCI to explain and review any draft report at BCI's request.

F. If Architect has been retained by BCI, under direction of the Legal Department of BCI, for the purpose of providing technical assistance to the Legal Department, in order that the Legal Department can formulate legal advice in connection with, or in anticipation of litigation, this section shall also apply. Unless BCI requests otherwise, any information and reports so generated by Architect shall be privileged and confidential. Architect shall include two cover pages with each copy of any draft report. Architect shall type on the first cover page on any correspondence to BCI the following heading:

DRAFT:
PRIVILEGED AND CONFIDENTIAL UNDER ATTORNEY-CLIENT PRIVILEGE AND WORK
PRODUCT DOCTRINE
PREPARED UNDER THE DIRECTION OF COUNSEL
DO NOT REPRODUCE OR DISTRIBUTE WITHOUT THE EXPRESS PERMISSION OF
DESIGNATED ATTORNEY

At BCI's direction, Architect shall prepare a final report that may or may not, at BCI's direction, contain the above privileged and confidential heading. Architect shall present draft and final copies of any reports prepared under the direction of legal counsel to the designated attorney.

14. Insurance

Any and all insurance and/or bonds that may be required under the Laws and Regulations of any governmental authority, including but not limited to, Worker's Compensation Insurance is, and shall be the sole responsibility of Architect.

A. Architect shall provide to BCI their current policy limits on an annual basis.

B. Without in any way limiting Architect's indemnification obligations as set forth in this Agreement, and prior to the commencement of any Services under a Contract, Architect shall take out and thereafter maintain in full force and effect the following insurance on an occurrence form basis unless otherwise stated:

1. Commercial General Liability Insurance (Bodily Injury and Property Damage), including the following supplementary coverage:
   a. Contractual Liability to cover liability assumed under this Agreement;
   b. Personal Injury Liability with the "employee" and "contractual" exclusions deleted;
   c. Product and Completed Operations Liability Insurance;
   d. Broad Form Property Damage Liability Insurance; and
   e. Explosion, collapse, and underground hazard coverage, if requested by BCI.

2. If use of a motor vehicle is required, Business Automobile Bodily Injury and Property Damage Liability Insurance. Such insurance shall extend to owned, if any, non-owned, and hired automobiles of Architect used in the performance of Services herein.

3. Unless BCI requests higher limits as set forth in this Agreement, the policy limits of liability for the insurance required above in B. 1), or B. 2), shall not be less than $1,000,000 per occurrence and $2,000,000 in the aggregate.

4. Employer's Liability Insurance with policy limits of not less than $100,000 for bodily injury each accident, $500,000 for bodily injury by disease policy limits, and $100,000 for bodily injury by disease-each employee.

5. Professional Liability Insurance for each Principal of Architect described in ARTICLE VII. Exhibit B. A. 2) with policy limits of not less than $1,000,000 per occurrence and $2,000,000 in the aggregate.

6. Aviation Liability Insurance if either Architect or its Consultants use a helicopter or other means of aviation in assisting in the performance of the Services, with policy limits of not less than $5,000,000.

C. Any Umbrella or Excess Liability Insurance shall provide that if an underlying aggregate of the above insurance is exhausted, the excess coverage shall drop down as primary insurance.

1. The Architect shall provide to BCI annually, evidence of the above insurance in the form acceptable to BCI. The document provided annually shall:

2. Name BCI and its Affiliates, and the directors, shareholders and employees of BCI and its Affiliates, as "additional insureds" in matters regarding the Services covered by this Agreement and/or any Contract;

3. Provide that said insurance is primary coverage with respect to all insureds;

4. Contain a Standard Cross Liability Endorsement which provides that the insurance applies separately to each insured against whom a claim is filed, and that the policies cover claims or suits by one insured against the other;

5. Contain a waiver of subrogation against BCI and its Affiliates;

6. The cancellation clause on the certificate of insurance will be amended to read as follows:

"SHOULD ANY OF THE ABOVE DESCRIBED POLICIES
BE CANCELED OR MATERIALLY CHANGED, THE ISSUING COMPANY WILL MAIL 30 DAYS
WRITTEN NOTICE TO THE CERTIFICATE HOLDER."

D. If requested by BCI at any time, Architect shall provide BCI with certification along with an endorsement/rider issued by a properly qualified representative of the insurer, specifying that Architect's insurance policy complies with this section, if requested by BCI, Architect shall provide to BCI a copy of any and all policies of insurance required pursuant to this Agreement and/or any Contract.

E. All insurance policies required shall be issued by companies admitted (licensed) to transact business in the State of California and who hold a current rating of B+ VII or better in the A.M. Best Key Rating Guide for Property and Casualty Insurance Companies.

F. Architect assigns to BCI and, as to BCI, waives all of its right, title, and interest in any and all claims, including but not limited to, claims of equitable, contractual, or statutory subrogation, which could be claimed against BCI arising from or in. connection with this Agreement for any and all losses, damages, or liability that would be covered by the insurance required herein, whether or not Architect maintains such coverage at the time the liability occurs. Architect revokes prior assignments or any of the aforesaid claims and agrees to notify its insurance carriers of the terms of this paragraph.

G. Architect's obligations to maintain the insurance required herein, and to provide evidence of it, shall survive for a period of ten (10) years beyond the termination, cancellation, or other expiration of this Agreement and/or any Contract; provided that if the Services includes asbestos abatement, such period shall be forty (40) years. If Architect's insurance coverage is made on "claims-made" forms, Architect agrees to maintain such insurance to cover liability arising out of or in connection with this Agreement and/or any Contract (i.e., adjust retroactive date, provide "tail coverage", etc.), and to provide to BCI evidence thereof for the period stated above in this paragraph.

H. At any time during the term of this Agreement and/or any Contract, BCI may require Architect to obtain and maintain in force, insurance with coverage or limits in addition to the foregoing, with charges as mutually agreed.

I. The Architect shall also require all Consultants who may enter upon the Project site to maintain the same insurance requirements listed above.

15. Liens

Architect shall not permit liens, encumbrances, or claims to be filed or asserted by others against BCI or BCI's property by reason of Architect's failure or alleged failure to pay for any labor performed or materials furnished pursuant to the terms of this Agreement and/or any Contract and shall protect, hold harmless, and indemnify BCI from and against all claims, liens, encumbrances, or claims of liens or encumbrances filed, or asserted, because of labor performed (or allegedly performed) or materials furnished (or allegedly furnished) hereunder, and shall be solely responsible for the payment of any such claim, and shall cause any such lien or encumbrance which may be filed, or asserted, to be immediately released and discharged of record. Upon failure of Architect to observe any of the provisions of this section, BCI at its option, may pay any or all of the claims involved and take such steps as it desires to cause such liens or encumbrances to be released. Architect shall thereupon, reimburse BCI for its reasonable Costs and payments. If Architect fails to reimburse BCI, then BCI may Withhold sufficient funds otherwise due Architect under this Agreement and/or Contract and may take such other action as is necessary to recover such costs and payments. BCI may recover from Architect all reasonable attorneys' fees expended by BCI in connection with any violation by Architect of the provisions of this section.

16. Limitation of liability

BCI will not be liable for consequential incidental, special or punitive damages, or for loss of revenue or profit in connection with the performance or failure to perform this Agreement regardless of whether such Liability arises from breach of contract, tort or any other theory of Liability.

17. Most favored Customer

Architect represents and warrants that all prices, benefits, warranties and other terms and conditions in this Agreement are and will continue to be during the term of this Agreement no less favorable than those currently being offered or which will be offered by Architect to any of its similarly situated customers. Architect shall review and have an officer of its company certify its compliance with this Section to BCI semi-annually. This certification shall be sent to BCI's representative listed under the Section, Notices.

18. Notices

Except as otherwise provided in this Agreement, or applicable Contract, all notices or communications here-under, shall be deemed to have been duly given when made in writing and either 1) delivered in person, 2) delivered by a recognized overnight delivery service, or 3) deposited in the United States Mail, postage prepaid, or 4) facsimile transmission, provided within 24 hours of transmission a copy is sent by any method set forth above in 1), 2) or 3), and addressed as follows:

Architect Notices:	Inland Architects, Inc.
	3130 N. Arbor Boulevard
	West Ontario, CA 91000
	Attn: Chief Architect
	Fax: 909-985-0230

Project Specific Notices:	Big Corp, Inc.
	100 E. Smith Ave
	Interior CA 91000
	Attn: Area Manager, Los Angeles Service Area
	Fax: 626-300-8001

Agreement Specific Notices:	Big Corp, Inc.
	2600 South 5th Street
	San Pablo, CA 91000
	Attn: Contract Manager-Alliance Services
	Fax: 909-876-9801

The address to which notices or other communications may be given by either Party may be changed by written notice given by such Party to the other pursuant to this section entitled "Notices."

19. Subcontractors

If Architect subcontracts any Services hereunder, all Services performed by a Subcontractor shall be deemed Services performed by Architect. Architect shall adopt and comply with the "Minority and Women Business Enterprise/Disabled Veteran Business Enterprise (M/ WBE-DVBE) Job Specific Subcontracting Plan" described in ARTICLE VII. Exhibit A-2.

If requested by BCI in a Contract or otherwise, the selection of any Subcontractor for performance of the Services required herein shall be subject to the written approval of BCI Representative. Each subcontractor selected for performing the services shall be satisfactorily experienced in their particular field with respect to both nature and magnitude of work as determined by BCI's Representative.

20. Waivers of Default

Unless otherwise expressly provided herein, no waiver by BCI of any provision hereof shall be deemed to have been made unless expressed in writing and signed by BCI. No delay or omission in the exercise of any right or remedy accruing to BCI upon any breach under this Agreement shall impair such right or remedy or be construed as a waiver of any such breach heretofore or thereafter Occurring. The waiver by BCI of any breach of any term, covenant or condition herein stated shall not be deemed to be a waiver of any other term, covenant, or condition.

21. Warranty

A. Architect warrants to BCI that any Services provided hereunder will be performed in a first-class professional manner, in strict compliance with the Specifications, and with the care, skill and diligence, and in accordance with the applicable standards, currently recognized in Architect's profession or industry. If Architect fails to meet applicable professional standards, Architect will, without additional compensation, promptly correct or revise any errors or deficiencies in the Services furnished hereunder.

B. The warranty period for Services shall be the longer of the warranty period stated in the Contract, the Specifications or one (1) year. The warranty period shall commence upon Acceptance.

C. Architect represents and warrants that:

1. There are no actions, suits, or proceedings pending or threatened, which will have a material adverse effect on Architect's ability to fulfill its obligations under this Agreement;

2. Architect will immediately notify BCI if, during the term of this Agreement, Architect becomes aware of any action, suit, or proceeding, pending or threatened, which may have a material adverse effect on Architect's ability to fulfill the obligations under this Agreement or any Contract;

3. Architect has all necessary skills, rights, financial resources, and authority to enter into this Agreement and related Contracts, including the authority to provide or license the Material or Services;

4. The Services will not infringe any patent, copyright, or other intellectual property;

5. No consent, approval, or withholding of objection is required from any entity, including any governmental authority with respect to the entering into or the performance of this Agreement or any Contract;

6. The Services will be provided free of any lien or encumbrance of any kind;

7. Architect will be fully responsible and liable for all acts, omissions, and Work performed by any of its representatives, including any subcontractor;

8. All representatives including subcontractors, will strictly comply with the provisions specified in this Agreement and any Contract; and,

9. Architect will strictly comply with the terms of this Agreement or Contract, including those specified in any Appendices thereto.

D. All warranties will survive inspection, acceptance, payment and use. These warranties will be in addition to all other warranties, express, implied or statutory. Architect will defend, indemnify and hold BCI harmless from and against all liabilities for a breach of these warranties.

E. If at any time during the warranty period for Services, BCI believes there is a breach of any warranty. BCI will notify Architect setting forth the nature of such claimed breach. Architect shall promptly investigate such claimed breach and shall either (i) provide Information satisfactory to BCI that no breach of warranty in fact occurred, or (ii) at no additional charge to BCI, promptly use its best efforts to take such action as may be required to correct such breach.

F. If a breach of warranty has not been corrected within a commercially reasonable time, or if two (2) or more breaches of warranty occur in any sixty (60) day period, BCI may Cancel the applicable Contract.

22. Signatures

IN WITNESS WHEREOF, the Parties hereto have caused this Agreement to be executed by their respective duly authorized representatives

BIG CORP, INC.,

By: _____

Printed Name: _____

Title: _____

Date Signed: _____

INLAND ARCHITECTS, INC.

By: _____

Printed Name: _____

Title: _____

Date Signed: _____

ARTICLE VII. EXHIBITS

Exhibit A – EXECUTIVE ORDERS AND ASSOCIATED REGULATIONS
Deleted

Exhibit B – SCHEDULE OF FEES AND LABOR RATES MATRIX A. FIXED PERCENTAGE FEE

MATRIX A. FIXED PERCENTAGE FEE

Cost of Construction	Fee for Work in Existing Data Centers (% of Construction $)	Fee for Work in Existing Admin Facilities (% of Construction $)	Fee for New Building Construction/Bldg Addition (% of Construction $)
$0–$250,000	7.95%	795%	7.95%
$250,001–$1,000,000	6.60%	6.60%	6.60%
$1,000,001–$5,000,000	6.50%	6.50%	6.50%
Over $5,000,000	6.40%	6.40%	6.40%

MATRIX B. OVERHEAD AND PROFIT (included in Fixed Percentage Fee)

Cost of Construction	Profit (% of Const $)	Over Head (% Of Const $)
$0–$250,000	7% of Design Fee 0.6% of Construction Cost	56.5% of Design Fee 4.5% of Construction Cost
$250,001–$1,000,000	7% of Design Fee 0.5% of Construction Cost	56.5% of Design Fee 3.7% of Construction Cost
$1,000,001–$5,000,000	7% of Design Fee 0.5% of Construction Cost	56.5% of Design Fee 3.7% of Construction Cost
Over $5,000,000	7% of Design Fee 0.4% of Construction Cost	56.5% of Design Fee 0.6% of Construction Cost

MATRIX C. PASS-THROUGH FEE

There will be no Pass-Through Fees on any Consultant fees, including Consultant fees such as, but not limited to, Acoustical Consultants, Geotechnical, Civil Engineers, etc. as may be required by BCI.

Cost of Work	Pass-thru fee (% of Consultant Fees)	Mechanical (% of Construction cost)	Electrical (% of Construction cost)	Structural (% of Construction cost)	Architectural (% of Construction cost)
$0–$250,000	0%	Not Applicable, Included within Matrix A	7.9%	Not Applicable, Included within Matrix A	
$250,001–$1,000,000	0%		6.60%		
$1,000,001–$5,000,000	0%		6.50%		
Over $5,000,000	0%		6.40%		

Note on Matrix C: Pass-Through fee should be separate from the mechanical/electrical/structural/architectural fees and includes profit margin.

MATRIX D. ARCHITECTURAL, MECHANICAL, ELECTRICAL and STRUCTURAL LABOR RATES for Inland Architects, Inc.

Job Title	Base Wage Rate ($/hr)	Multiplier	Billing Rate ($/hr)	Overtime Hrs Multiplier
Principal	56.50	No Multiplier	155.00	N/A
Project Manager	54.55	2.75	150.00	N/A
Project Architect	34.65	2.75	95.30	N/A
Project Engineer	38.85	2.75	106.85	N/A
CAD Operator	17.75	2.75	48.80	N/A
Clerical	23.09	2.75	63.50	N/A

EXHIBIT B-1: – REIMBURSABLE EXPENSES

Architect shall be reimbursed at Actual Cost to Architect, including Consultants, for the following items connected with the Project except as otherwise provided in the Agreement:

1. Reproduction, postage and handling of Drawings and Specification for bidding and Construction.
2. Plan check, permit and filing fees required by this Agreement, ordinance, code or law
3. Interstate telephone calls, telegrams or special messages.

4.   Expenses in connection with authorized out-of-town (out-of-town is defined as a Project site located outside the city where the employee's main office is located) travel including mileage reimbursed at the rate allowed by the Internal Revenue Service for the use of a personal vehicle or the cost of commercial transportation only in excess of individual's daily commute cost, parking cost (excluding parking related to visiting Architect/Consultant place of business), meals at actual cost not to exceed $25. per meal with a daily maximum of $30.00, and long distance telephone calls. Meals shall be reimbursed only if the destination (i.e., Project location) is located over 200 miles (one-way) from the employee's main office.

The following items require prior written authorization by BCI's Representative:

5.   Special Services, such as but not limited to, renderings, models, photographs and mock-ups.

6.   Other Services requested by BCI's Representative not listed above.

7.   The total of all reimbursable expenses shall not exceed the limit indicated for such reimbursable expenses on the Contract without the prior written approval of BCI's Representative in writing and in accordance with SBC's Schedule of Approvals prior to start of Work. A current copy of BCI's Schedule of Approvals will be provided to Architect by BCI on request BCI will not pay for Reimbursables that are not authorized per this Agreement.

CPSIA information can be obtained
at www.ICGtesting.com
Printed in the USA
FSOW03n0218240217
31152FS